Physical Chemistry
A Guided Inquiry

Atoms, Molecules, and Spectroscopy

Richard S. Moog
Franklin & Marshall College

James N. Spencer
Franklin & Marshall College

John J. Farrell
Franklin & Marshall College

Houghton Mifflin Company Boston New York

Publisher: Charles Hartford
Executive Editor: Richard Stratton
Assistant Editor: Danielle Richardson
Editorial Assistant: Rosemary Mack
Senior Project Editor: Fred Burns
Manufacturing Manager: Florence Cadran
Marketing Manager: Katherine Greig

Printed in the U.S.A.

ISBN: 0-618-30854-7
1 2 3 4 5 6 7 8 9—POO—07 06 05 04 03

To the Instructor

Physical Chemistry: A Guided Inquiry: Atoms, Molecules, and Spectroscopy is not a textbook. This book is meant to be used in class as a *guided inquiry*. Much research has shown that more learning takes place when the student is actively engaged, and ideas and concepts are developed by the student, rather than being presented by an *authority* — a textbook or an instructor[1]. The *ChemActivities* presented here are structured so that information is presented to the reader in some form (an equation, a table, figures, written prose, etc.) followed by a series of Critical Thinking Questions which lead the student to the development of a particular concept or idea. Whenever possible, data are presented *before* a theoretical explanation, and the Critical Thinking Questions lead the learner through the thought processes which results in the construction of a particular theoretical model. This is what makes this book a *guided inquiry*. We have tried to mimic the scientific process as much as possible throughout this book. Students are often asked to make predictions based on the model that has been developed up to that point, and then further data or information is provided which can be compared to the prediction. It is important that these predictions be made BEFORE proceeding to get the full appreciation and benefit of this way of thinking. In this way, models can be confirmed, refined, or refuted, using the paradigm of the scientific method. The philosophical and pedagogic basis for this approach is described in a recent article.[2]

In addition to the student edition of *Physical Chemistry: A Guided Inquiry: Atoms, Molecules, and Spectroscopy*, we have written several other documents to assist you in your endeavor to make the students better learners:

- The Instructor's Edition of *Physical Chemistry: A Guided Inquiry: Atoms, Molecules, and Spectroscopy*. Available online rather than in printed form, this book is identical to the student edition except that it has answers to all of the focus questions and critical thinking questions (CTQs) in shaded boxes directly under the CTQs. Our answers are not the only answers or even the best answers. Rather, they serve to alert you as to the type of response we are expecting.
- The Solutions Manual for *Physical Chemistry: A Guided Inquiry: Atoms, Molecules, and Spectroscopy*. This book contains worked-out solutions for all of the exercises and problems at the end of each ChemActivity. The solutions manual may be purchased by students.
- The Instructor's Guide for *Physical Chemistry: A Guided Inquiry: Atoms, Molecules, and Spectroscopy*. This online guide gives suggestions for how to organize the course, how to assign groups and roles, how to interact with groups, and how to respond to student questions. It also contains a table that correlates each ChemActivity to chapters in several other physical chemistry textbooks, the major concepts covered in each ChemActivity, the approximate time for each ChemActivity, and sample quizzes.

---

[1] Johnson, D. W. ; Johnson, R. T. *Cooperative Learning and Achievement.* In Sharon, S. (Ed.), *Cooperative Learning: Theory and Research*, pp 23-37, New York: Praeger.

[2] Spencer, J. N. *J. Chem. Educ.* **1999**, *76*, 566.

For information on accessing the online supplements (the Instructor's Edition and Instructor's Guide) please contact your Houghton Mifflin sales representative. You can find out who your representative is by visiting Houghton Mifflin's web site at college.hmco.com.

Richard S. Moog	richard.moog@fandm.edu	Franklin & Marshall College
James N. Spencer	james.spencer@fandm.edu	Chemistry Department
John J. Farrell		Lancaster, PA 17604

To the Student

Physical chemistry is a subdiscipline of chemistry that encompasses a quantitative study of the physical properties of chemicals and chemical reactions. For about a century the standard topics of physical chemistry have been thermodynamics (heat flow) and kinetics (rates of chemical reactions). These topics are generally concerned with the macroscopic properties of chemicals and chemical reactions. During the past 40 years or so, physical chemists have placed an increasing emphasis on understanding chemicals and chemical reactions at the molecular level—analysis of virtually one atom or molecule rather than a large collection of molecules. On the theoretical level this analysis is called quantum mechanics. On the experimental level this analysis is typically achieved by some form of spectroscopy.

This book about physical chemistry is *not* a textbook. This book is *not* a study guide. This book is a *guided inquiry*. Specifically, this book is a collection of group activities (each group has three or four students) that are to be accomplished in the presence of a mentor (instructor). Each group activity has one or more *models* (data, prose, or figures that represent the core of some chemical concept) followed by a series of *critical thinking questions* (CTQs). Systematically working through the CTQs in groups is essential for three basic reasons:

- Explaining concepts to other members of your group not only helps *their* understanding, it broadens *your* understanding. Instructors often have an exceptional understanding of the material they teach. One of the reasons for this depth of understanding is that teachers are constantly explaining concepts and exchanging ideas. Research has shown that this sort of verbal communication is a very important aspect of the learning process. Furthermore, it is often the case that someone who has just learned a concept is a better communicator of the concept to a novice than someone who is very familiar with the concept.
- Learning to ask questions that clearly and concisely describe what you do not understand is an important skill (not only in this and other courses but in all aspects of your life). It is a skill that improves with practice. When you do not receive the answer you expected from the other members of your group you may realize that the failing was in your question. You will learn how to ask better questions from your mistakes, from your mentor, from other members of your groups, and from reading the CTQs in the book.
- Groups (teams) have become essential to identifying, defining, and solving problems in our society. It is important that we learn how to be active and productive members of a group. If a member of your group is not contributing, it is your responsibility to help that member to become more productive. If a member of your group is over-contributing (thereby slowing the progress of the group) it is your responsibility to help that member to become more productive. Furthermore, as a member of the group you will have a role to play (manager, recorder, technician, and so on). Each role has a function important to the success of the group. Understanding the roles and dynamics of a group and how

to change the dynamics of a group is a skill that can be transferred to many real-life situations.

We have found the use of these methods to be a more effective learning strategy than the traditional lecture, and the vast majority of our students have agreed. We hope that you will take ownership of your learning and that you will develop skills for lifelong learning. No one else can do it for you. We wish you well in this undertaking.

If you have any suggestions on how to improve this book, please write to us.

Richard S. Moog richard.moog@fandm.edu Franklin & Marshall College
James N. Spencer james.spencer@fandm.edu Chemistry Department
John J. Farrell Lancaster, PA 17604

Acknowledgments

This book is the result of the innumerable interactions that we have had with a large number of stimulating and thoughtful people.

- Thanks to the following faculty members who have used some of our preliminary materials and have provided helpful feedback: Renee Cole (Central State University); Jeffrey Kovac (University of Tennessee); Ken Morton (Carson-Newman College); Susan Phillips (University of Pennsylvania); Marty Perry (Ouachita Baptist University); Clayton Spencer (Illinois College); George Shalhoub (La Salle University).

- Thanks to Richard Stratton, Houghton Mifflin Company, who had the vision and courage to bring a new educational pedagogy to college-level students. Thanks to Danielle Richardson, Katherine Greig, and Alexandra Shaw, all of Houghton Mifflin Company, for their help and encouragement throughout the entire process.

- Special thanks to Dan Apple, Pacific Crest Software, for taking us to this untravelled path and for pushing during the first two years of the journey. The Pacific Crest Teaching Institute we attended provided us with the insights and inspiration to convert our classroom into a fully student-centered environment.

- We also appreciate the support and encouragement of the many members of the Middle Atlantic Discovery Chemistry Project, who have provided us with an opportunity to discuss our ideas with interested, stimulating, and dedicated colleagues.

- Thanks to Carol Strausser, Franklin & Marshall College, for typing the photo-ready copy and for having sufficient patience to work with us through the editing and reediting.

- A great debt of thanks is due our students in Physical Chemistry at Franklin & Marshall College these past five years. Their enthusiasm for this approach, patience with our errors, and helpful and insightful comments have inspired us to continue to develop as instructors, and have helped us to improve these materials immeasurably.

Contents

ChemActivity **1**

The Energies of Molecules

(How do Molecules Move?)

Assume for a moment that you could make a movie of a molecule whizzing through space. From a distance, the molecule would appear as a single object moving in a straight line. The molecule would only change direction when it encountered another molecule. Then, as when two billiard balls collide, each molecule would change direction (according to the laws of conservation of energy and momentum) and move in a straight line until the next collision. As you zoomed in on the molecule, however, you would see that what appeared to be a single object was, in fact, a collection of objects (atoms). Furthermore, the atoms within the molecule would not be moving in a straight line; the motion of the atoms would appear to be much more complicated than the straight-line motion of the molecule as a whole—and somewhat erratic. As you zoomed in for a close-up of an atom, you would see that what appeared to be a single object (an atom in this case) was, in fact, a collection of many objects (a nucleus and many electrons). The electrons would move much faster than the molecule as a whole and much faster than the nuclei.

The purpose of this ChemActivity is to differentiate conceptually the rather complex motion of the various particles in a molecule. This separation of motion leads to four different types of molecular energy: translational; rotational; vibrational; electronic. This differentiation of motion will help us understand chemical transformations and the absorption and emission of electromagnetic radiation (spectroscopy).

Model 1: The Energy of a Particle can be Subdivided According to Different Types.

A useful approximation is that the various motions that a particle has can be considered individually and independently, and the associated energies can be summed to give the total energy of the particle. That is, the presence (or absence) of one type of motion has no effect on the other types of motion.

Table 1. Energy Values for Various Particles

Particle	MW (g/mole)	Translational Energy 25°C (kJ/mole)	Rotational Energy 25°C (kJ/mole)	Vibrational Energy 25°C (kJ/mole)	Electronic Energy 25°C (kJ/mole)
H(g)	1.01	3.72	0	0	–1,312
C(g)	12.01	3.72	0	0	–99,386
O(g)	16.00	3.72	0	0	–197,182
F(g)	19.00	3.72	0	0	–262,040
Si(g)	28.09	3.72	0	0	–761,102
Ar(g)	39.95	3.72	0	0	–1,389,178
CO(g)	28.01	3.72	2.48	12.8	–297,654
HF(g)	20.01	3.72	2.48	23.7	–263,922
HCl(g)	36.46	3.72	2.48	17.3	–1,211,909

Critical Thinking Questions

1. What correlation is there, if any, between the MW of a particle and

   a) translational energy?

   b) rotational energy?

   c) vibrational energy?

   d) electronic energy?

2. What do all particles with no rotational energy have in common?

3. What do all particles with no vibrational energy have in common?

Information

The center of mass of a diatomic molecule is defined as shown in Figure 1.

Figure 1. Definition of the Center of Mass of a Diatomic Molecule.

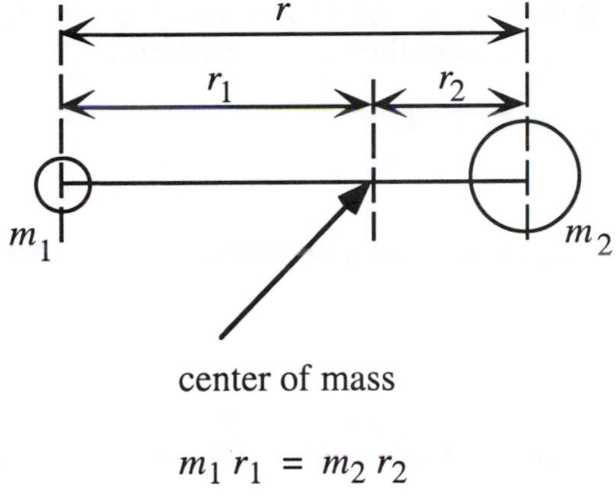

$$m_1 r_1 = m_2 r_2$$

Model 2: Translational Energy.

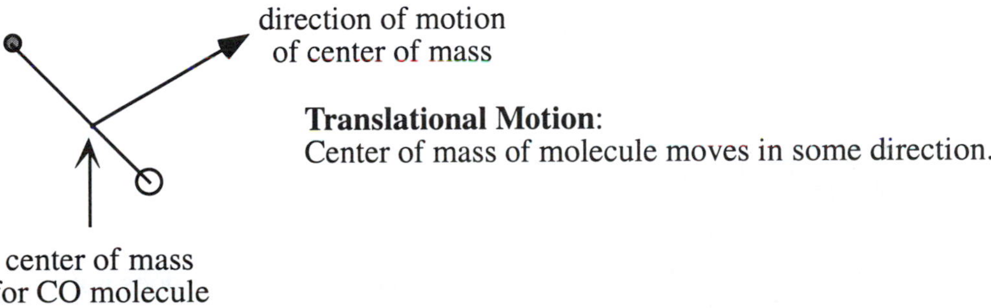

The **translational energy** of a molecule can be thought of as $\frac{1}{2}\ mv^2$, where all of the mass of the molecule, m, is located at the center of mass of the molecule.

Critical Thinking Questions

4. Is the center of mass of a diatomic molecule closer to the heavier atom or the lighter atom?

5. Which atom is the carbon atom in Model 2? Explain your answer.

6. The internuclear distance (bond length) in CO is 113 pm. The triple-bond radius of a carbon atom is 60 pm, and the triple-bond radius of an oxygen atom is 53 pm. Does this mean that the center of mass of CO is located on the straight line connecting the two nuclei and 60 pm from the carbon nucleus?

Exercises

1. The molecular mass of ^{12}C is exactly 12 g/mole. The molecular mass of ^{16}O is 15.995 g/mole. Determine the mass of one carbon-12 atom (to four significant figures). Determine the mass of one oxygen-16 atom (to four significant figures).

2. Determine the location of the center of mass of $^{12}C^{16}O$ to four significant figures given that the internuclear distance is 113.1 pm.

Model 3: Rotational Energy.

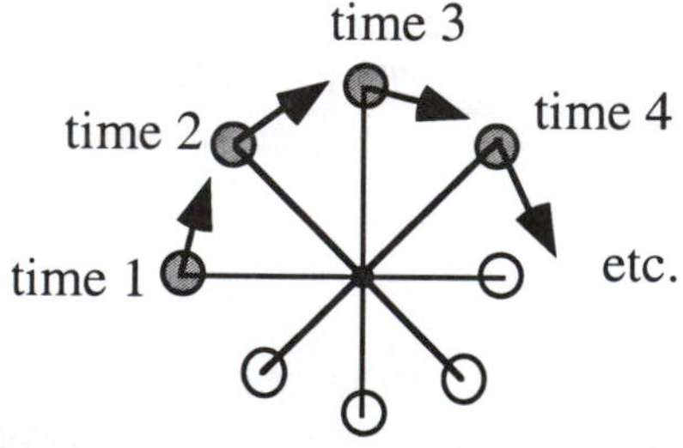

Rotational Motion:
Center of mass is fixed.
Distance between nuclei does not change.
Nuclei rotate around the center of mass.

The energy associated with the motion of two or more nuclei around the center of mass of a molecule is the **rotational energy**.

Critical Thinking Question

7. Why is the rotational energy of C(g), O(g), and Ar(g) equal to zero?

Model 4: Vibrational Energy.

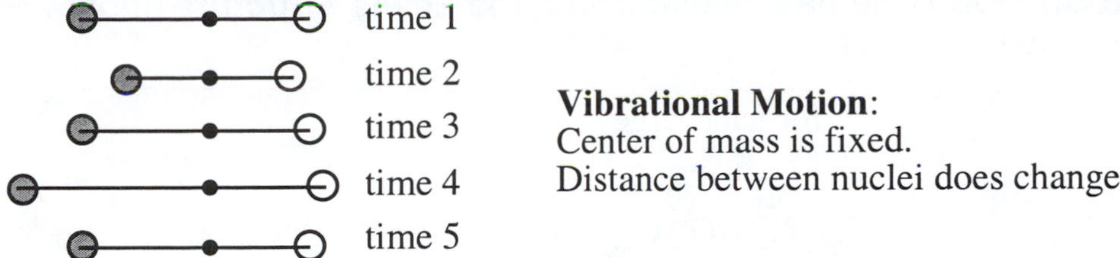

Vibrational Motion:
Center of mass is fixed.
Distance between nuclei does change.

The energy associated with the change of the distance between nuclei is the **vibrational energy**.

Critical Thinking Question

8. Why is the vibrational energy of C(g), O(g), and Ar(g) equal to zero?

Model 5: Electronic Energy.

The **electronic energy** is the sum of the following:

- The coulombic interaction of all of the electrons in the molecule with all of the nuclei in the molecule (all of these are attractive forces and negative energies).
- The coulombic interaction of all of the electrons in the molecule with each other (all of these are repulsive forces and positive energies).
- The coulombic interaction of all of the nuclei in the molecule with each other (all of these are repulsive forces and are positive energies).
- The magnetic interaction of all of the particles (electrons and nuclei) in the molecule.
- The kinetic energy of the electrons.

The motion of the center of mass of the molecule, the rotation of the nuclei around the center of mass, and the motion of the nuclei with respect to each other are not part of the electronic energy.

Critical Thinking Questions

9. What or whose law mathematically describes the potential energy of interaction between a positively charged particle and a negatively charged particle?

10. Use this law (answer to CTQ 9) to show that the potential energy of interaction between a positive and a negative charge is a negative quantity.

Information

Figure 1. Carbon monoxide has 14 electrons; the Lewis structure shows 10 of the electrons.

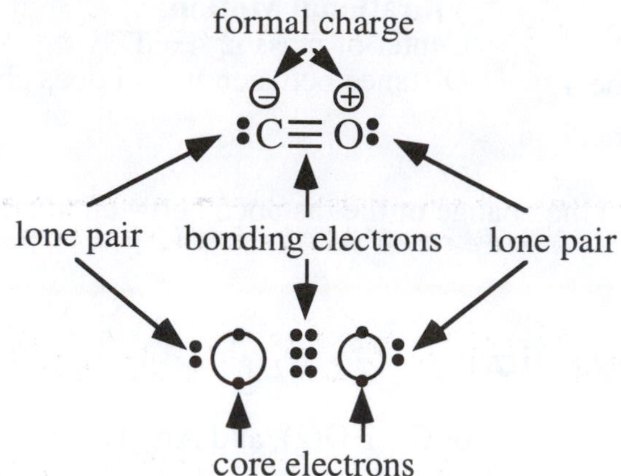

The electronic energy, U_{elec}, of CO can be thought of as the ΔU for the following process:

$$14\ e^- \text{(stationary)} + C^{6+} \text{(stationary)} + O^{8+} \text{(stationary)} = CO \text{(stationary)} \quad (1)$$

Equation (1) can be written as the following thermodynamic sequence:

Process	ΔU (kJ/mole)	Quantity Description[a]
$C^{6+}(g) + 1\ e^-(g) = C^{5+}(g)$	–47,276	($-IE_6$ of carbon)
$C^{5+}(g) + 1\ e^-(g) = C^{4+}(g)$	–37,829	($-IE_5$ of carbon)
$C^{4+}(g) + 1\ e^-(g) = C^{3+}(g)$	–6,223	($-IE_4$ of carbon)
$C^{3+}(g) + 1\ e^-(g) = C^{2+}(g)$	–4,620	($-IE_3$ of carbon)
$C^{2+}(g) + 1\ e^-(g) = C^{1+}(g)$	–2,352	($-IE_2$ of carbon)
$C^{1+}(g) + 1\ e^-(g) = C(g)$	–1,086	($-IE_1$ of carbon)
$O^{8+}(g) + 1\ e^-(g) = O^{7+}(g)$	–84,074	($-IE_8$ of oxygen)
$O^{7+}(g) + 1\ e^-(g) = O^{6+}(g)$	–71,322	($-IE_7$ of oxygen)
$O^{6+}(g) + 1\ e^-(g) = O^{5+}(g)$	–13,326	($-IE_6$ of oxygen)
$O^{5+}(g) + 1\ e^-(g) = O^{4+}(g)$	–10,989	($-IE_5$ of oxygen)
$O^{4+}(g) + 1\ e^-(g) = O^{3+}(g)$	–7,469	($-IE_4$ of oxygen)
$O^{3+}(g) + 1\ e^-(g) = O^{2+}(g)$	–5,300	($-IE_3$ of oxygen)
$O^{2+}(g) + 1\ e^-(g) = O^{1+}(g)$	–3,388	($-IE_2$ of oxygen)
$O^{1+}(g) + 1\ e^-(g) = O(g)$	–1,314	($-IE_1$ of oxygen)
$C(g) + O(g) = CO(g)$	–1,086	(–CO bond energy)[b]
$14\ e^-(g) + C^{6+}(g) + O^{8+}(g) = CO(g)$	–297,654	(electronic energy of CO)

[a] IE_i is the ith ionization energy

[b] The normal CO bond energy of 1074 kJ/mole has been corrected by 12 kJ/mole to account for the rotational, vibrational, and translational energy change between CO(g) and the separated atoms.

Critical Thinking Questions

11. Why is the value of i in IE_i equal to 5 in the following equation?

$$C^{4+}(g) = C^{5+}(g) + 1\ e^{-}(g)$$

12. Is energy required or released for the reaction in CTQ 11? Why?

13. Why is IE_6 for carbon more positive than IE_1 for carbon?

14. Why is IE_1 for oxygen more positive than IE_1 for carbon?

15. IE_1 for oxygen is 1,314 kJ/mole. The IE_i for oxygen increase gradually to 13,326 kJ/mole for IE_6. Why is IE_7 for oxygen so large, 71,322 kJ/mole?

16. Is energy required or released when C and O form CO? Explain.

17. According to the thermodynamic sequence given for equation (1), what is the value of the electronic energy of CO?

18. According to the thermodynamic sequence given for equation (1), what is the value of the bond energy of CO?

19. What percent of the electronic energy of CO is due to the CO energy released when C(g) and O(g) react to form CO(g)? (Calculate a numerical value.)

20. The electronic energy for CO is a very large negative number. Therefore, which is stronger: the attractive forces between the electrons and the nuclei, or the repulsive forces (electron/electron and nuclei/nuclei)? Explain.

Exercises

3. Explain why the electronic energy of a molecule is a negative quantity whereas the translational, rotational, and vibrational energies are positive quantities.

4. Explain why, in general, the electronic energy of an atom or molecule becomes more negative as the MW increases.

5. Use the electronic energies of H(g) and O(g), see Model 1, to estimate the electronic energy of OH(g). Correct this value by taking into account the bond energy of OH, 459 kJ/mole. What fraction of the electronic energy of OH is due to the OH bond energy (calculate a numerical value)?

6. Predict which atom in each of the following pairs has the more negative electronic energy. Explain your reasoning.
   a. Be or Mg b. Si or K c. Ne or Na

7. Predict which molecule in each of the following pairs has the more negative electronic energy. Explain your reasoning.
   a. I_2 or F_2 b. CH_4 or CF_4 c. H_2O or H_2S

Problems

1. a) Find the appropriate ionization values in the Handbook of Chemistry and Physics and calculate U_{elec} for Li(g).

   b) Ignore the bond energies and calculate U_{elec} for LiH(g) and LiF(g). [Hint: see Model 1.]

2. Using grammatically correct English sentences, define the center of mass of a diatomic molecule with atoms of mass m_1 and m_2.

ChemActivity 2

Quantum Mechanics
(Time Independent)

The fundamental equation of **classical mechanics** is $f = ma$, or more exactly, $f = \frac{dp}{dt}$. There is no derivation of this equation. There is no proof of this equation. Scientists use this equation because it is useful.

The fundamental equation of **quantum mechanics** is $\hat{H}\Psi = \varepsilon\Psi$. This is often called the **time independent Schrödinger equation**. There is no derivation of this equation. There is no proof of this equation. Scientists use this equation because it is useful.

The application of quantum mechanics for time-independent systems (such as the structure of atoms and molecules in their ground states) is based on several postulates.[1] As stated above, there is no derivation or proof of these postulates. They are assumed to be true and are then used to describe physical systems. They have proved to be very useful, so we will use them also!

Postulate I. The state of a system is completely defined by Ψ, a mathematical function of the coordinates of the components of the system. This function must be finite, continuous, and single valued. Ψ contains all of the information about the system, and is called a **wavefunction** or an **eigenfunction**.

Postulate II. Every dynamical variable (or physical observable) is represented by a corresponding linear operator.

Postulate III. When a dynamical variable A is measured (without experimental error), there are only certain possible values that may be obtained. These values are the eigenvalues a of the operator A as given by

$$\hat{A}\phi = a\phi \quad .$$

Here ϕ is an eigenfunction of the operator $\hat{A}$ that represents the dynamical variable A.

Postulate IV. The wavefunction Ψ for the state of a system is provided as a solution of the equation

$$\hat{H}\Psi = \varepsilon\Psi \quad .$$

where $\hat{H}$ is the operator for the total energy of the system, also know as the **Hamiltonian operator**.

---

[1]For time-dependent phenomena, a more complicated form of the Schrödinger equation is required. In these materials, we will not be concerned with this more complicated formulation.

Because you are probably not familiar with many of these terms, or the quantum mechanical way of thinking about things, we will begin with some explanations.

- Ψ is a mathematical description of a single particle or a system of particles. It is a function of the position of the particle (or the positions of the particles). Thus, Ψ for a single particle inside a 3-dimensional box would be $\Psi(x,y,z)$.
- Ψ^2 is a probability function.[2] Ψ (and Ψ^2) is zero in any region where the probability of finding the particle is zero. Ψ can be positive or negative (or imaginary), but Ψ^2 is always positive and real. Ψ can never be infinite.
- ε is the numerical value of the energy of the particle (system).
- The concept of an operator is important in quantum mechanics. Operators are represented by symbols such as $\hat{A}$. The hat above the A designates a mathematical operation or a series of mathematical operations. For example, the "2" in the quantity x^2 is a mathematical operator; in this case, it says to multiply x by x. The symbol $\frac{d}{dx}$ is also an operator; it says take the derivative with respect to x of any function to the right of the operator. The square root sign, as in $\sqrt{5}$. is also an operator; it says find a quantity such that the quantity squared is 5. The Hamiltonian operator $\hat{H}$ is a set of mathematical operations such that when applied to the wavefunction that describes the particle (system) the value of the energy multiplied by the wavefunction is produced.
- An operator is a linear operator if it obeys the distribution law with respect to addition of functions. That is, an operator $\hat{A}$ is linear if

  $$\hat{A}(\Psi_1 + \Psi_2) = \hat{A}\Psi_1 + \hat{A}\Psi_2.$$

  For example, the operator $\frac{d}{dx}$ is linear; $\sqrt{}$ is not.

[2]This is only rigorously correct for real functions Ψ. We will address this subtlety in CA 3.

Model 1: Solving a Physical Problem with Quantum Mechanics: a Step-wise Approach.

1. Write the classical energy for the particle (system).

   Explanation. The classical energy, U, is a sum of the kinetic energy, T, and the potential energy, V.

$$U = T + V \qquad (1)$$

2. Convert velocity, v, to linear momentum, p. (Sometimes it is more convenient to convert velocity to angular momentum, L; these cases will be treated later.)

   Explanation. The kinetic energy of a single particle is

$$\frac{1}{2} mv^2$$

   Convert to linear momentum as follows:

$$\frac{1}{2} mv^2 = \frac{1}{2}\frac{m^2v^2}{m} = \frac{p^2}{2m}$$

3. Substitute the appropriate operators for all terms in the total energy, $T + V$. This is an application of Postulate II.

   The fundamental operator is the operator for linear momentum in one direction, $\hat{p}_x$.

$$\hat{p}_x = \frac{\hbar}{i}\frac{d}{dx}$$

   where $\hbar = \dfrac{h}{2\pi}$, h is Planck's constant, and $i = \sqrt{-1}$.

   The three-dimensional counterpart is

$$\hat{p} = \frac{\hbar}{i}\left(\frac{\partial}{\partial x} + \frac{\partial}{\partial y} + \frac{\partial}{\partial z}\right)$$

   The operator for a number is simply the number. The operator for a physical constant is simply the constant (the operator for the mass of an electron is m_e or 9.109×10^{-31} kg; the operator for the charge on an electron is $-e$; the operator for the nuclear charge is Ze). The operator for a coordinate is simply the coordinate (the operator for ϕ is ϕ).

   All other operators are derived from the momentum operator. For example, the operator for kinetic energy in the x direction, $\hat{T}_x$ is derived as follows:

$$T_x = \frac{1}{2} mv_x^2 = \frac{p_x^2}{2m}$$

$$\hat{T}_x = \frac{1}{2\hat{m}}\hat{p}_x^2 = \frac{1}{2m}\hat{p}_x^2\,\hat{p}_x^2 = \frac{1}{2m}\frac{\hbar}{i}\frac{d}{dx}\frac{\hbar}{i}\frac{d}{dx} = -\frac{\hbar^2}{2m}\frac{d^2}{dx^2}$$

Table 1 gives some quantities and the associated operators.

Table 1. Operators for Physical Quantities.

Quantity	Operator Symbol	Operator
N_A (Avogadro's number)	$\hat{N}_A$	N_A (or 6.022×10^{23})
m_e (mass of electron)	$\hat{m}_e$	m_e (or 9.109×10^{-31} kg)
x	$\hat{x}$	x
z	$\hat{z}$	z
ϕ	$\hat{\phi}$	ϕ
r	$\hat{r}$	r
p_x	$\hat{p}_x$	$\frac{\hbar}{i}\frac{\partial}{\partial x}$
T_x	$\hat{T}_x$	$-\frac{\hbar^2}{2m}\frac{\partial^2}{\partial x^2}$
T (Cartesian coordinates)	$\hat{T}$	$-\frac{\hbar^2}{2m}\left(\frac{\partial^2}{\partial x^2} + \frac{\partial^2}{\partial y^2} + \frac{\partial^2}{\partial z^2}\right)$
T (spherical coordinates)	$\hat{T}$	$-\frac{\hbar^2}{2m}\left\{\frac{1}{r^2}\frac{\partial}{\partial r}\left(r^2\frac{\partial}{\partial r}\right) + \frac{1}{\sin\theta}\left[\frac{\partial}{\partial\theta}(\sin\theta)\frac{\partial}{\partial\theta}\right] + \frac{1}{\sin^2\theta}\frac{\partial^2}{\partial\phi^2}\right\}$
L_x (Cartesian coordinates)	$\hat{L}_x$	$\frac{\hbar}{i}\left(y\frac{\partial}{\partial z} - z\frac{\partial}{\partial y}\right)$
L_x (spherical coordinates)	$\hat{L}_x$	$\frac{\hbar}{i}\left(-\sin\phi\frac{\partial}{\partial\theta} - \cot\theta\cos\phi\frac{\partial}{\partial\phi}\right)$
L_y (Cartesian coordinates)	$\hat{L}_y$	$\frac{\hbar}{i}\left(z\frac{\partial}{\partial x} - x\frac{\partial}{\partial z}\right)$
L_y (spherical coordinates)	$\hat{L}_y$	$\frac{\hbar}{i}\left(\cos\phi\frac{\partial}{\partial\theta} - \cot\theta\sin\phi\frac{\partial}{\partial\phi}\right)$
L_z (Cartesian coordinates)	$\hat{L}_z$	$\frac{\hbar}{i}\left(x\frac{\partial}{\partial y} - y\frac{\partial}{\partial x}\right)$
L_z (spherical coordinates)	$\hat{L}_z$	$\frac{\hbar}{i}\frac{\partial}{\partial\phi}$
L^2 (spherical coordinates)	$\hat{L}^2$	$-\hbar^2\left\{\frac{1}{\sin\theta}\left[\frac{\partial}{\partial\theta}(\sin\theta)\frac{\partial}{\partial\theta}\right] + \frac{1}{\sin^2\theta}\frac{\partial^2}{\partial\phi^2}\right\}$

4. Let $\hat{U} = \hat{T} + \hat{V} = \hat{H}$ Recall that $\hat{H}$ is the **Hamiltonion operator**. (2)

5. Now it is time to apply Postulate IV. Right-hand multiply $\hat{H}$ by Ψ and set equal to $\varepsilon\Psi$. Ψ is called the **wavefunction**. At this point in the process the form of Ψ is not known. Ψ is, however, some mathematical function that contains all of the pertinent information about the physical system under consideration.

   $$\hat{H}\Psi = \varepsilon\Psi \quad (3)$$

   Equation (3) is called the Schrödinger equation. This type of equation is frequently encountered in mathematics and science. The general form is:

   $\hat{A}\phi = a\phi$ where a is some constant.

   In general, the function ϕ is called the **eigenfunction** of the operator $\hat{A}$, and a is called the **eigenvalue**. Note that Postulate III indicates that the eigenvalues a are the only possible values whch can be obtained by measurement of the variable A.

6. Solve equation (3), a differential equation, for Ψ. The idea is to find the function (or functions) which, when substituted into equation (3) for Ψ, make the equation true. (That is, the left side equals the right side!)

   Explanation. Normally, solution of equation (3) is not trivial. The solutions are typically a set of polynomials. Often, the solutions are named after the person who first solved the differential equation or who discovered the polynomials: the Laguerre polynomials ; the Hermite polynomials. The solutions always depend on the boundary conditions. That is, the particle might be constrained to a particular region of space (a line, a circle, sphere, a disk, and so on). It is not necessary to know the value of ε to solve the differential equation; it is necessary to know or assume that ε is constant.

7. If there are any parameters in Ψ, solve for these parameters by normalization or by application of boundary conditions.

   Explanation. Solution of a differential equation usually results in one or more parameters that depend on the unique character of the problem. For example, often Ψ is the mathematical description of an electron, and Ψ^2 is interpreted as being the probability for the electron.

   Thus, the probability of finding the electron between x_1 and x_2, y_1 and y_2, z_1 and z_2 is equal to

   $$\int_{x_1}^{x_2}\int_{y_1}^{y_2}\int_{z_1}^{z_2} \Psi^2\, dz\, dy\, dx$$

   Of course, the probability of finding the electron in the range $x = -\infty$ to $x = \infty$, $y = -\infty$ to $y = \infty$, $z = -\infty$ to $z = \infty$, must be one:

$$\int_{-\infty}^{\infty}\int_{-\infty}^{\infty}\int_{-\infty}^{\infty} \Psi^2 \, dz \, dy \, dx = 1 \tag{4}$$

If Ψ contains a parameter, it can now be determined such that equation (4) is valid.

8. Solve for ε.

   Explanation. Now that Ψ is known, the energy can be obtained from equation (3).

Critical Thinking Questions

1. Boltzmann's constant, k, is called the universal gas constant per molecule because it is equal to the universal gas constant, R, divided by Avogadro's constant, N_A. What is the operator for Boltzmann's constant?

2. What is the expression for the translational energy of a particle moving in a straight line in the y direction?

   According to Table 1, what is the operator for translational energy in the y direction, $\hat{T}_y$?

3. At some point in space Ψ^2 has one of the following values. Which one is correct? Why?

   $\Psi^2 = 1.57 \times 10^{-5}\ \text{pm}^{-3}$
   $\Psi^2 = \infty\ \text{pm}^{-3}$
   $\Psi^2 = -2.64 \times 10^{-4}\ \text{pm}^{-3}$

4. Of what significance is the unit pm^{-3}?

Exercises

1. The gravitational potential energy of a proton and an electron separated by a distance r is given by:

$$V = G\frac{(m_p)(m_e)}{r}$$

where m_p is the proton mass, m_e is the electron mass, and G is the Newtonian constant of gravitation.

What is the operator for this potential energy?

2. Recall that the z-component of angular momentum is given by:

$$L_z = x\,p_y - y\,p_x$$

Show that the operator for L_z in Table 1 is correct.

3. According to Table 1, what is the operator for:

$$\frac{1}{2}\,mv_z^2 + \frac{1}{2}\,kz^2$$

given that z is a coordinate, m is the mass of a particle, and k is some constant?

4. The potential energy of a proton and an electron separated by a distance r is given by Coulomb's law:

$$\frac{(e)(-e)/4\pi\varepsilon_o}{r} = \frac{-e^2/4\pi\varepsilon_o}{r}$$

where e is the magnitude of the charge on an electron and ε_o is the vacuum permittivity (a fundamental constant).

What is the operator for this potential energy?

5. The classical expression for the kinetic energy of a particle under certain conditions is:

$$T = \frac{L^2}{2I}$$

where L^2 is the square of the magnitude of the angular momentum and I is the moment of inertia, mr^2.

What is the operator for this kinetic energy?

Problems

1. Evaluate $g = \hat{Q} f$, where $\hat{Q}$ and f are given below:

	$\hat{Q}$	f
(a)	$\sqrt{\ }$	x^6
(b)	$\frac{d^2}{dx^2}$	e^{-ax}
(c)	$\left(\frac{\partial^2}{\partial x^2} + \frac{\partial^2}{\partial y^2} + \frac{\partial^2}{\partial z^2}\right)$	$x^2y^3z^5$

2. Show that $e^{\alpha x}$ is an eigenfunction of the operator d^2/dx^2. What is the eigenvalue?

3. Show that $\sin \beta x$ is an eigenfunction of the operator d^2/dx^2. What is the eigenvalue?

ChemActivity **3**

Translational Energy (I)

(Where, oh where, has my little particle gone?)

The classical expression for the kinetic energy of a moving particle (the translational energy) is:

$$T = \frac{1}{2} mv^2$$

In classical mechanics there are no restrictions on the kinetic energy of the moving particle—the particle can have any kinetic energy between zero and infinite joules. In quantum mechanics, however, we will find that there are restrictions on the kinetic energy the particle can have.

We will begin by considering a particle moving in a straight line—on the x axis. The particle will be constrained to a particular region of the x axis—between $x = 0$ and $x = a$ (a is some value on the x axis). We do not need to propose why the particle is constrained to this region. It could be that there is a physical barrier at $x = 0$ and at $x = a$ that cannot be penetrated. Or, it could be that there is some mysterious field that doesn't allow the particle to leave the region between $x = 0$ and $x = a$. Most particles are constrained to some region in space because of electric, magnetic, or gravitational fields. Here, we want to focus on the translational energy of the particle confined to a region on a line. For simplicity, we will assume that there are no forces (electric, magnetic, coulombic) acting on the particle in the allowed region (between $x = 0$ and $x = a$). This model is called the **particle-on-a-line**.

Model 1: A Particle-on-a-Line.

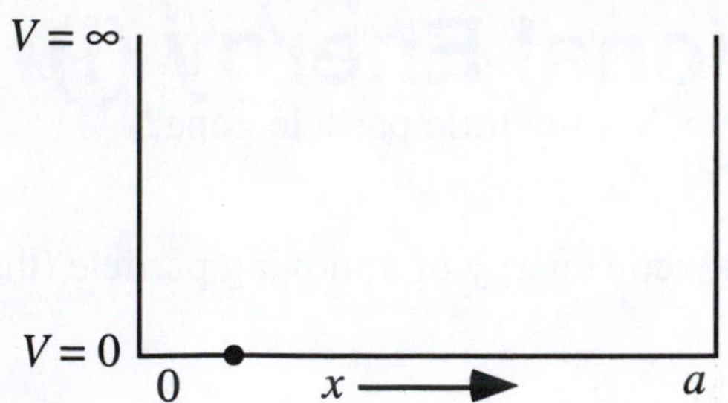

For $x < 0$ and $x > a$:

$V = \infty$

$\Psi(x) = 0$ (The particle is constrained to the interval $0< x <a$) (1)

For $x > 0$ and $x < a$:

V = 0 (There are no forces acting on the particle in this region.)

$$U = T + V = \frac{1}{2} mv_x^2 + 0 = \frac{p_x^2}{2m} \qquad 0< x <a \qquad (2)$$

$$\hat{H} = -\frac{\hbar^2}{2m}\frac{d^2}{dx^2} \qquad 0< x <a \qquad (3)$$

$$\hat{H}\Psi = -\frac{\hbar^2}{2m}\frac{d^2\Psi}{dx^2} = \varepsilon\Psi \qquad 0< x <a \qquad (4)$$

$$\Psi_{n_x} = \sqrt{\frac{2}{a}}\cos\frac{n_x\pi x}{a} \qquad n_x = 1, 2, 3, \ldots \qquad (5)$$

or

$$\Psi_{n_x} = \sqrt{\frac{2}{a}}\sin\frac{n_x\pi x}{a} \qquad n_x = 1, 2, 3, \ldots \qquad (6)$$

or

$$\Psi_{n_x} = \sqrt{\frac{2}{a}}\, e^{-in_x\pi x/a} \qquad n_x = 1, 2, 3, \ldots \qquad (7)$$

$$\varepsilon_{n_x} = \frac{\pi^2 n_x^2 \hbar^2}{2ma^2} = \frac{n_x^2 h^2}{8ma^2} \qquad n_x = 1, 2, 3, \ldots \qquad (8)$$

Critical Thinking Questions

1. Why is $\Psi_{n_x} = 0$ for $x < 0$ and $x > a$?

2. Equation (5) is a solution to equation (4):

$$\frac{d}{dx}\left(\sqrt{\frac{2}{a}}\cos\frac{n_x\pi x}{a}\right) = -\frac{n_x\pi}{a}\sqrt{\frac{2}{a}}\sin\frac{n_x\pi x}{a}$$

$$\frac{d^2}{dx^2}\left(\sqrt{\frac{2}{a}}\cos\frac{n_x\pi x}{a}\right) = \frac{d}{dx}\left(-\frac{n_x\pi}{a}\sqrt{\frac{2}{a}}\sin\frac{n_x\pi x}{a}\right) = -\frac{n_x^2\pi^2}{a^2}\sqrt{\frac{2}{a}}\cos\frac{n_x\pi x}{a}$$

$$-\frac{\hbar^2}{2m}\frac{d^2\Psi}{dx^2} = -\frac{\hbar^2}{2m}\frac{d^2}{dx^2}\left(\sqrt{\frac{2}{a}}\cos\frac{n_x\pi x}{a}\right) = \frac{n_x^2h^2}{8ma^2}\sqrt{\frac{2}{a}}\cos\frac{n_x\pi x}{a} = \frac{n_x^2h^2}{8ma^2}\Psi$$

$$-\frac{\hbar^2}{2m}\frac{d^2\Psi}{dx^2} = \frac{n_x^2\,h^2}{8ma^2}\Psi = \varepsilon\Psi$$

What must the value of ε be for equation (4) to be true?

3. Show that equation (6) is a solution to equation (4), and indicate the restrictions on the possible values for ε.

4. Ψ_{n_x} is equal to zero at x values less than zero and at x values greater than a because the probability of finding the particle at those values is zero. Recall that, according to Postulate I, Ψ_{n_x} must be continuous and single-valued; therefore, $\Psi_{n_x} = 0$ at $x = 0$ and at $x = a$. Two of the three equations, (5), (6), and (7), do not obey this condition. Which Ψ_{n_x} does obey this condition: equation (5), (6), or (7)?

5. According to equation (8), what happens to the energy of the particle as the value of n_x increases?

6. According to equation (8), what happens to the value of ε_1 (ε with $n_x = 1$) if the length of the line is increased to $2a$?

7. A sketch of Ψ_1 ($n_x = 1$), Ψ_2 ($n_x = 2$), and Ψ_3 ($n_x = 3$) is given below. A node is defined as a location where Ψ is equal to zero (other than at the boundaries). Derive an equation that gives the number of nodes as a function of n_x.

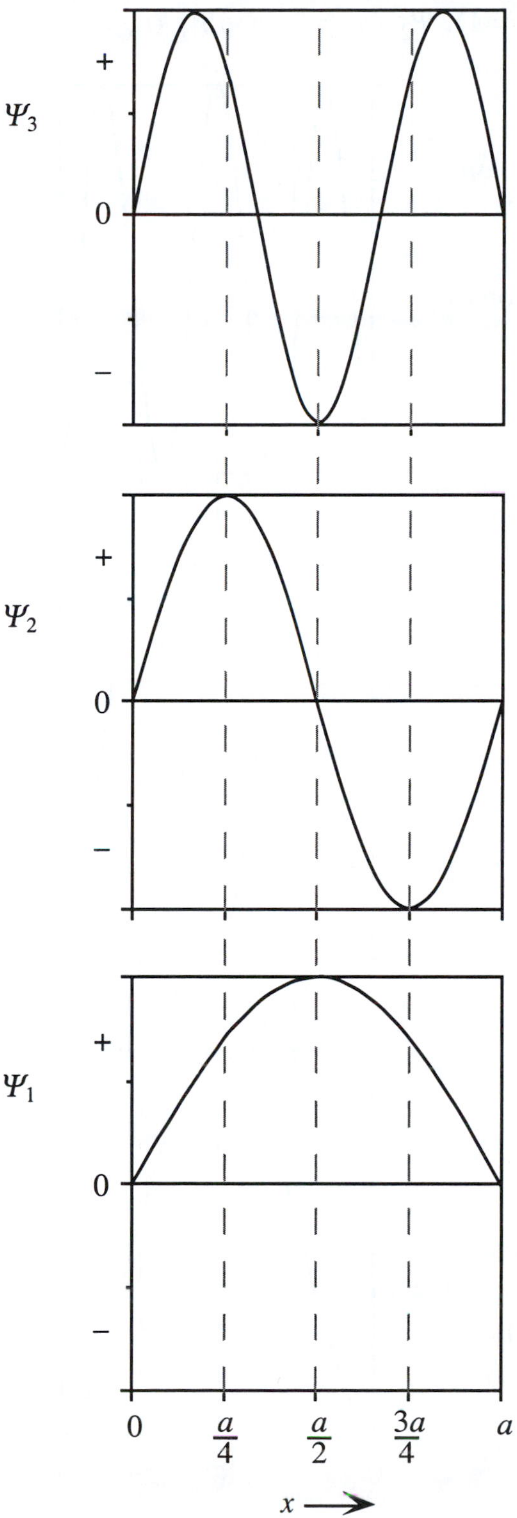

8. What is the qualitative relationship between the number of nodes and the energy for a particle-on-a-line?

9. A sketch of Ψ_1^2 ($n_x = 1$), Ψ_2^2 ($n_x = 2$), and Ψ_3^2 ($n_x = 3$) is given below.

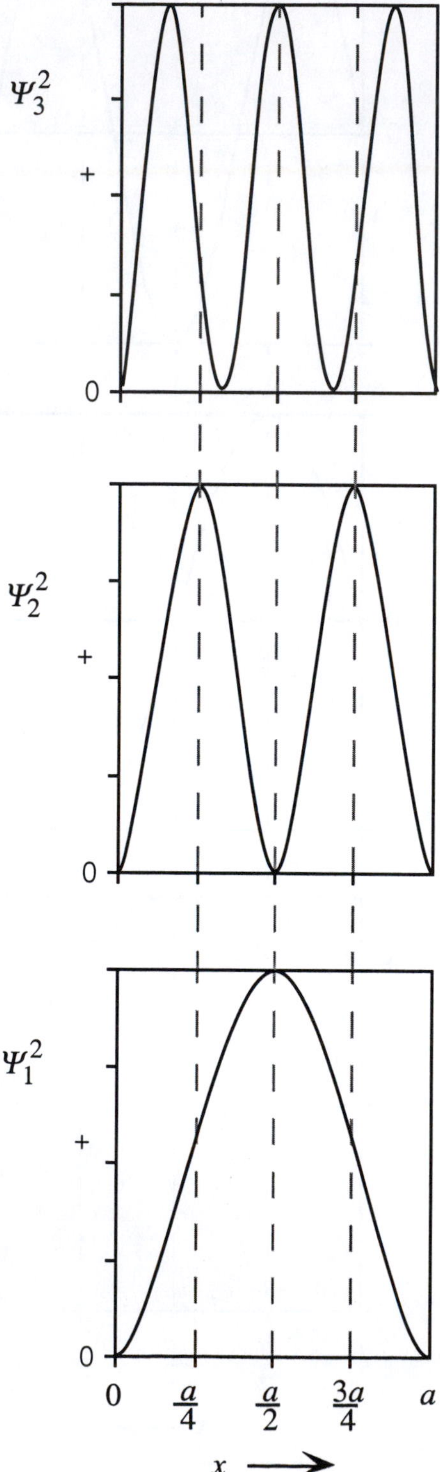

a. For a particle described by $n_x = 1$, is it more likely to find the particle at $a/4$, at $a/2$, or at $3a/4$? Explain.

b. For a particle described by $n_x = 2$, is it more likely to find the particle at $a/4$, at $a/2$, or at $3a/4$? Explain.

10. Given that

$$\Psi_{n_x}^2 = \frac{2}{a}\sin^2\left(\frac{n_x \pi x}{a}\right)$$

At what values of $\frac{n_x \pi x}{a}$ between $x = 0$ and $x = a$ will $\Psi_{n_x}^2$ be a maximum? Use this information to determine the three most probable location (values of x) of a particle described by $n_x = 3$.

11. Predict the average value of x for a particle described by $n_x = 1$.

12. Predict the average value of x for a particle described by $n_x = 2$.

13. Predict the average value of x for a particle described by $n_x = 27$.

14. A sketch of Ψ_1 and Ψ_2 is given below. Without explicitly doing the integration, show that $\int_0^a \Psi_1 \Psi_2 \, dx = 0$. [Hint: this integral is equal to $\sum_{x=0}^{x=a} \Psi_1(x)\Psi_2(x)$.]

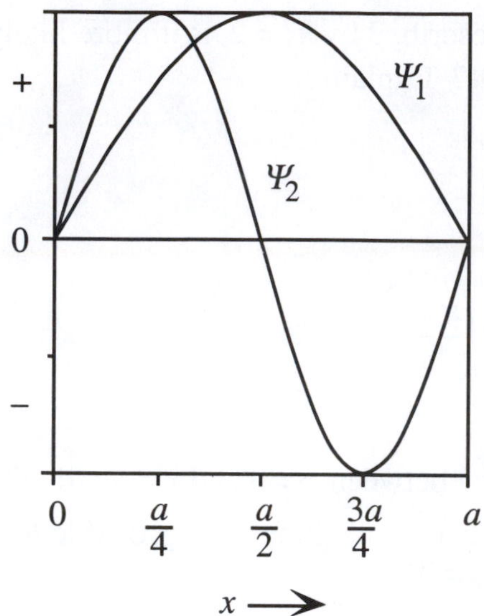

Information

Ψ^* is the complex conjugate of Ψ. The complex conjugate is obtained by replacing i by –i everywhere it appears in the function, where $i = \sqrt{-1}$. If Ψ is a real function (that is, i does not appear as part of the function) then $\Psi^* = \Psi$.

Generally, wave functions will be **orthogonal**: $\int \Psi_i^* \Psi_j \, d\tau = 0$;

for real functions: $\int \Psi_i \Psi_j \, d\tau = 0$

where Ψ_i and Ψ_j are two different wave functions that can describe a system, and $d\tau$ is the volume element appropriate for integration over all space. In the case of the particle-on-a-line: $d\tau = dx$ and the integral spans $x = 0$ to $x = a$.

Generally, wave functions will be **normalized**: $\int \Psi_i^* \Psi_j \, d\tau = 1$;

for real functions: $\int \Psi_i \Psi_j \, d\tau = 1$

where Ψ_i is a wave function that describes a system. The integral is set equal to one because there must be unit probability of finding the particle somewhere.

Many of the systems we will be dealing with involve real functions Ψ.

Exercises

1. Given equation (2), show that the Hamiltonian operator in equation (3) is correct. [Hint: use Table 1 of CA 2 on Quantum Mechanics.]

2. Recall that $\Psi_{n_x}^2 = \frac{2}{a} \sin^2 \frac{n_x \pi x}{a}$ $\quad n_x = 1, 2, 3, \ldots$
   Find one location between $x = 0$ and $x = a$ where Ψ_6^2 = is a maximum.

3. Explain how the following statement is related to the answer to CTQ 6.

   An electron in a π-system, such as 1,3 butadiene or benzene, is spread out (delocalized) over a large region of space (compared to an electron bound only to a single atom or to an electron confined to two atoms in a normal covalent bond). It is well-known that the electrons in a π-system are at a lower energy than in a normal π bond (the energy is said to be lowered by **delocalization**).

4. Show, by integration, that Ψ_1 and Ψ_2 are orthogonal.

5. Show, by integration, that Ψ_1 is normalized.

6. Show that equation (7) is a solution to equation (4).

Information

When a photon is emitted or absorbed by a molecule (or atom), the molecule changes from one energy state (designated by a set of quantum numbers) to another energy state (designated by a new set of quantum numbers). The difference in energy between the two states is equal to the energy of the photon. This is known as the Bohr frequency rule:

$$E_{\text{photon}} = h\nu = \frac{hc}{\lambda} = \varepsilon_{\text{higher}} - \varepsilon_{\text{lower}}$$

where $\varepsilon_{\text{higher}}$ and $\varepsilon_{\text{lower}}$ are the energy values for the for the higher and lower energy states of the molecule or atom.

Model 3: The Average Value/The Average Grade on an Exam.

Grade	Number of Students Who Received Grade (NS)	Number of Students × Grade (NS × Grade)	Relative probability of Receiving a Grade (PG)	Relative probability of Receiving a Grade × Grade (PG × Grade)
100	0	0	0	0
99	0	0	0	0
98	0	0	0	0
⋮	0	0	0	0
90	1	90	0.20	18
89	0	0	0	0
88	0	0	0	0
⋮	0	0	0	0
80	2	160	0.40	32
79	0	0	0	0
⋮	0	0	0	0
70	1	70	0.20	14
69	0	0	0	0
⋮	0	0	0	0
1	0	0	0	0
Sum	4	320	0.80	64

There are two ways to obtain the average.

Method 1: $$\text{average grade} = \frac{\text{sum of (NS} \times \text{Grade)}}{\text{total NS}} = 320/4 = 80$$

Method 2: $$\text{average grade} = \frac{\text{sum of (PG} \times \text{Grade)}}{\text{sum of PG}} = 64/0.80 = 80$$

Critical Thinking Questions

15. Which method, 1 or 2, is the usual way to determine the average grade on an exam?

16. If the relative probabilities of receiving 90, 80, and 70 were given as 0.15, 0.30, and 0.15 respectively, would this change the average grade using method 2?

Information

The average value equation in quantum mechanics is

$$<q> = \frac{\int \Psi_i^* \hat{q} \Psi_j \, d\tau}{\int \Psi_i^* \Psi_j \, d\tau} \tag{9}$$

Here, q is the quantity for which the average value is wanted and $\hat{q}$ is the operator for q. This equation is similar to Method 2, on the previous page. An integral is used rather than a summation because wave functions are continuous, unlike grades on an exam. The numerator of equation (9) is the sum of the following: the value of q at every point in space times the probability that the particle will be at that point. That is, if

$$\hat{q}\,\Psi_\text{i} = q_\text{i}\Psi_\text{i} \,,$$

$$\text{then} \quad \Psi_\text{i}\,\hat{q}\,\Psi_\text{i} = q_i\Psi_\text{i}\,\Psi_\text{i} = q_i\Psi_\text{i}^2$$

The beauty of equation (9) is that it is valid even when Ψ_i is not a valid wavefunction for q. That is, equation (9) will result in a average value for q even when

$$\hat{q}\,\Psi_\text{i} \neq q_\text{i}\Psi_\text{i} \,.$$

The denominator is simply the sum of the probabilities. Usually the wave function will be normalized, $\int \Psi_i \Psi_j \, d\tau = 1$, but equation (9) covers the cases where Ψ is not normalized.

Exercises

7. Use the average value equation to find the average value of x for a particle-on-a-line described by $n_x = 1$. Does this integral give a value for $<x>$ in agreement with the answer given in CTQ 11?

8. Use the average value equation to find the average value of x for a particle-on-a-line described by $n_x = 2$. Does this integral give a value for $<x>$ in agreement with the answer given in CTQ 12?

Problems

1. The particle-on-a-line problem is often used as a model for the π-electrons in conjugated molecules. Assume that the π-electrons of 1,3 butadiene move along a straight line of 578 pm (two C=C bond lengths of 135 pm, one C–C bond length of 154 pm, plus an addition one-half of a C–C bond length at the two terminal carbon atoms: 77 + 135 + 154 + 135 + 77 = 578 pm). The energy level diagram is shown below.

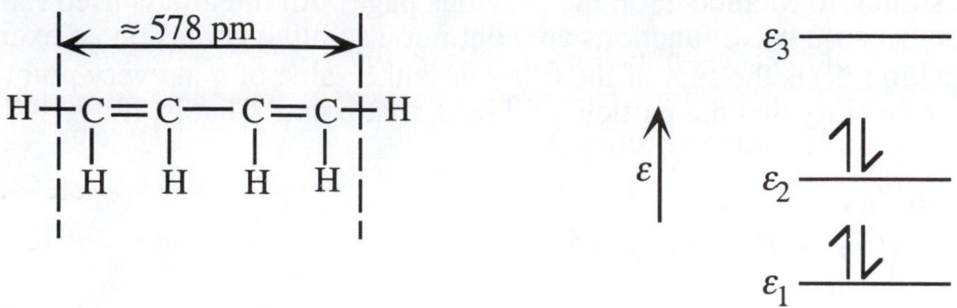

a) Why are there four electrons in the energy level diagram?
b) Why are only two electrons placed in ε_1?
c) What is the energy of an electron in ε_2 (in joules)?
d) At what energy (in joules) is the ε_3 level?
e) What energy is required to promote an electron from ε_2 to ε_3 (in joules)?
f) If this energy is supplied by a photon, what is the wavelength of the photon?
g) Compare this wavelength to the experimentally observed wavelength of 217 nm.

ChemActivity 4

Translational Energy (II)

(Should 3-D glasses be supplied with this activity?)

We have seen that the quantum mechanical solution to the particle-on-a-line problem (translational motion of a confined particle) has led to several important concepts:

- **Quantization** of energy—not all energies are **allowed**.
- A set of **wave functions**, Ψ_i, to describe the particle—each wave function is characterized by a **quantum number**.
- A set of **probability functions**, Ψ_i^2 (or $\Psi_i^* \Psi_i$), to describe the particle—the particle described by Ψ_i is not uniformly distributed in the space that the particle can occupy.
- The greater the number of **nodes** for the wave function that describes the particle, the higher the energy of the particle.
- The set of wave functions that describe the particle are **orthogonal**.
- Generally, the set of wave functions that describe the particle are **normalized**.

Now, we extend the one-dimensional translational problem to three dimensions. This model is called the **particle-in-a-box**.

Model 1: A Particle-in-a-Box.

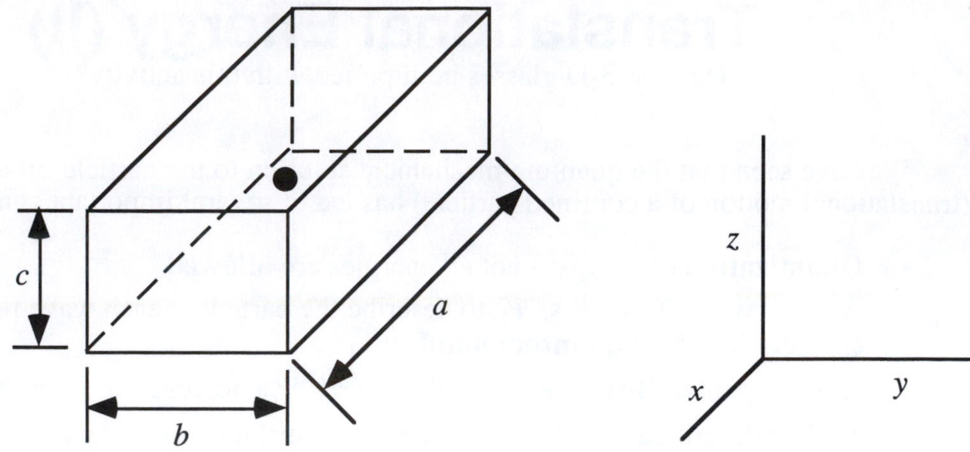

$$E = \frac{1}{2}mv_x^2 + \frac{1}{2}mv_y^2 + \frac{1}{2}mv_z^2 = \frac{p_x^2}{2m} + \frac{p_y^2}{2m} + \frac{p_z^2}{2m} \tag{1}$$

$$\hat{H} = -\frac{\hbar^2}{2m}\left(\frac{\partial^2}{\partial x^2} + \frac{\partial^2}{\partial y^2} + \frac{\partial^2}{\partial z^2}\right) = -\frac{\hbar^2}{2m}\nabla^2 \tag{2}$$

where $\nabla^2 = \left(\frac{\partial^2}{\partial x^2} + \frac{\partial^2}{\partial y^2} + \frac{\partial^2}{\partial z^2}\right)$ = the Laplacian operator

$$\hat{H}\,\Psi = -\frac{\hbar^2}{2m}\left(\frac{\partial^2 \Psi}{\partial x^2} + \frac{\partial^2 \Psi}{\partial y^2} + \frac{\partial^2 \Psi}{\partial z^2}\right) = \varepsilon\Psi \tag{3}$$

$$\Psi_{n_x,n_y,n_z} = \mathrm{X}_{n_x}\mathrm{Y}_{n_y}\mathrm{Z}_{n_z} \tag{4}$$

$$\mathrm{X}_{n_x} = \sqrt{\frac{2}{a}}\,\sin\frac{n_x \pi x}{a} \qquad n_x = 1, 2, 3, \ldots \tag{5}$$

$$\mathrm{Y}_{n_y} = \sqrt{\frac{2}{b}}\,\sin\frac{n_y \pi y}{b} \qquad n_y = 1, 2, 3, \ldots \tag{6}$$

$$\mathrm{Z}_{n_z} = \sqrt{\frac{2}{c}}\,\sin\frac{n_z \pi z}{c} \qquad n_z = 1, 2, 3, \ldots \tag{7}$$

$$\varepsilon_{n_x,n_y,n_z} = \varepsilon_{n_x} + \varepsilon_{n_y} + \varepsilon_{n_z} = \frac{n_x^2 h^2}{8ma^2} + \frac{n_y^2 h^2}{8mb^2} + \frac{n_z^2 h^2}{8mc^2} \tag{8}$$

Critical Thinking Questions

1. In a cubic box, $a = b = c$. The wavefunction is:

$$\Psi_{n_x,n_y,n_z} = X_{n_x}Y_{n_y}Z_{n_z} = \left(\frac{2}{a}\right)^{3/2} \sin\frac{n_x\pi x}{a} \sin\frac{n_y\pi y}{a} \sin\frac{n_z\pi z}{a}$$

   If a particle is described by $n_x = n_y = n_z = 1$, where is the most probable location of the particle in the x direction?

2. In a cubic box, $a = b = c$. If a particle is described by $n_x = n_y = n_z = 1$, where is the most probable location of the particle in the y direction?

3. In a cubic box, $a = b = c$. If a particle is described by $n_x = n_y = n_z = 1$, where is the most probable location of the particle in the z direction?

4. In a cubic box, $a = b = c$. If a particle is described by $n_x = n_y = n_z = 1$, where is the most probable location of the particle?

5. In a cubic box, $a = b = c$. Which state is the lower energy state in a cubic box, $\Psi_{1,1,1}$ or $\Psi_{2,2,2}$?

6. Which state is the lower energy in a cubic box, $\Psi_{2,2,2}$ or $\Psi_{3,1,1}$?

7. Which state is the lower energy in a cubic box, $\Psi_{2,1,2}$ or $\Psi_{2,2,1}$?

8. Is there another state that has the same energy in a cubic box as $\Psi_{2,1,2}$ and $\Psi_{2,2,1}$.

Information

It is very common for two or more states with different quantum numbers to have the same energy. The states are said to be **degenerate**. In a cubic box, above, for example, the energy level that has the sum of the n_i^2 equal to nine is said to be three-fold degenerate. The underlying cause of this degeneracy is the geometry and orientation of the motion. All directions are the same in a cubic box. One particle might be described by $\Psi_{2,1,2}$ and another particle might be described as $\Psi_{2,2,1}$. Both particles would have the same total energy, but they would have different energies in the *y* and *z* directions.

Critical Thinking Questions

9. In a rectangular box where $a = b$ and $c = 2a$, which state is the lower energy , $\Psi_{2,1,2}$ or $\Psi_{2,2,1}$? [Note that the degeneracy can be removed by making one of the dimensions of the box different than the other two. Also note that all of the degeneracy is removed when $a \neq b \neq c$.]

10. In a rectangular box where $a = b$ and $c = 2a$, which state is the lower energy, $\Psi_{2,1,2}$ or $\Psi_{1,2,2}$?

Exercises

1. Determine the degeneracy of the (3,3,3) energy level in a cubic box. [Hint: the answer is greater than three.]

2. Use the information given in Model 1 of ChemActivity 3 and Model 1 of ChemActivity 4 to find the expressions for Ψ and ε for a particle-in-a-square (particle in a 2D box).

3. Find the most probable location (one) for a particle-in-a-square described by $\Psi_{1,1}$. [Notation used is Ψ_{n_x,n_y}.]

4. Find the most probable locations (four) for a particle-in-a-square described by $\Psi_{2,2}$.

5. Find the most probable locations (two) for a particle-in-a-square described by $\Psi_{1,2}$.

6. Calculate the energy, in joules, of a CO molecule described by $\Psi_{2,5,3}$ in a box of dimensions $x = 10$ cm, $y = 10$ cm, and $z = 10$ cm.

7. According to Model 1 of CA 1 on Energies of Molecules, the translational energy of one mole of CO molecules at 25°C is 3.72 kJ. What is the average translational energy of one CO molecule at 25°C? For a box of dimensions $x = 10$ cm, $y = 10$ cm, and $z = 10$ cm, let $n_x = n_y = n_z = n$, and calculate the value of n for a CO molecule which has the average translational energy at 25°C.

Problems

1. What are the units for Ψ and Ψ^2 of a particle-in-a-box. [Hint: see equations (4)-(7).] Explain why the units for Ψ^2 make sense (recall that Ψ^2 is a probability function).

2. Show that $\Phi =$ (sin ax) (sin by) (sin cz) is an eigenfunction of the Laplacian operator,

$$\nabla^2 = \left(\frac{\partial^2}{\partial x^2} + \frac{\partial^2}{\partial y^2} + \frac{\partial^2}{\partial z^2}\right)$$

ChemActivity **5**

Molecular Vibration

(How fast do molecules vibrate?)

The quantum mechanical solution to the particle-in-a-box problem (three dimensional translational motion of a confined particle) has led to several important concepts:

- **Quantization** of energy—not all energies are **allowed**.
- A set of **wave functions,** Ψ_i, to describe the particle—each wave function is characterized by a set of **quantum numbers**. We have only investigated two cases, but it appears that the number of quantum numbers used to describe the particle is equal to the number of dimensions (one quantum number for the one-dimensional problem and three quantum numbers for the three-dimensional problem.
- More than one **energy state** can exist at a particular **energy level**. Each energy state has a unique set of quantum numbers. The number of energy states that have the same energy is the **degeneracy** of the energy level.
- A set of **probability functions**, Ψ_i^2 (or $\Psi_i^* \Psi_i$), to describe the particle—the particle described by Ψ_i is not uniformly distributed in the space that the particle can occupy.
- The greater the number of **nodes** for thewave function that describes the particle, the higher the energy of the particle.
- The set of wave functions that describe the particle are **orthogonal**.
- Generally, the set of wave functions that describe the particle are **normalized**.

Now, we attempt another one-dimensional problem—the vibrational motion of a diatomic molecule. The basic model for the vibrational motion is a ball and spring—this model is called the **harmonic oscillator**.

Model 1: Hooke's Law and the Harmonic Oscillator.

Consider a ball of mass μ attached to a spring as shown below. As the mass moves back and forth the spring compresses and expands. In its simplest manifestation, it is assumed that the spring exerts a restoring force that is proportional to the displacement from the equilibrium length of the spring. This is known as Hooke's law.

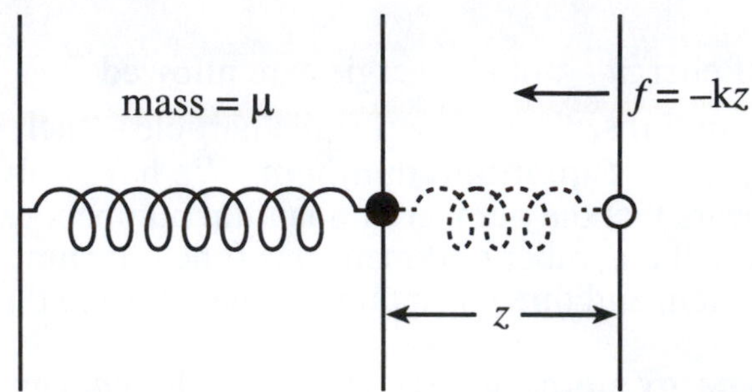

$$f = -kz \qquad \text{Hooke's Law} \tag{1}$$

where z is the displacement from the spring's rest position (at equilibrium).

Because $V = -\int f\,dz$,

$$V = -\int_0^z (-kz)\,dz = \frac{1}{2}kz^2 \tag{2}$$

We know that the kinetic energy of the particle is given by:

$$T = \frac{1}{2}\mu v^2 \tag{3}$$

$$U = T + V = \frac{1}{2}\mu v_z^2 + \frac{1}{2}kz^2 = \frac{p_z^2}{2\mu} + \frac{1}{2}kz^2 \tag{4}$$

$$\hat{H} = -\frac{\hbar^2}{2\mu}\frac{d^2}{dz^2} + \frac{1}{2}kz^2 \tag{5}$$

$$\hat{H}\Psi = -\frac{\hbar^2}{2\mu}\frac{d^2\Psi}{dz^2} + \frac{1}{2}kz^2\Psi = \varepsilon\Psi \tag{6}$$

This differential equation has the same form as one that was first solved by Hermite. The normalized solutions are

$$\Psi_v = N_v H_v(x)\, e^{-az^2/2} \tag{7}$$

where N_v is the normalization constant for Ψ_v, $H_v(x)$ are the Hermite polynomials (see Appendix A.3 on Polynomials), and $a = \frac{2\pi}{h}\sqrt{k\mu}$.

The first two normalized solutions to equation (4) are:

$$\Psi_0 = \left(\frac{a}{\pi}\right)^{1/4} e^{\frac{-az^2}{2}} \qquad (8)$$

$$\Psi_1 = \left(\frac{4a^3}{\pi}\right)^{1/4} z\, e^{\frac{-az^2}{2}} \qquad (9)$$

The vibrational energies, ε_υ, are given by:

$$\varepsilon_\upsilon = (\upsilon + \tfrac{1}{2})\, \hbar \sqrt{\frac{k}{\mu}} \qquad \upsilon = 0, 1, 2, \ldots \qquad (10)$$

Critical Thinking Questions

1. According to equation (10), what happens to the vibrational energy of the particle if the vibrational quantum number υ is increased (holding μ and k constant)?

2. According to equation (10), what happens to the vibrational energy of the particle if the mass, μ, is replaced by a heavier mass (holding υ and k constant)?

3. According to equation (10), what happens to the vibrational energy of the particle if the force constant, k, is replaced by a larger force constant (holding υ and μ constant)?

Exercises

1. Given equation (4), show that the Hamiltonian operator, equation (5), is correct. [Hint: see CA 2 on Quantum Mechanics.]

2. The Schrödinger equation for the harmonic oscillator, equation (6), is

$$-\frac{\hbar^2}{2\mu}\frac{d^2\Psi}{dz^2} + \frac{1}{2}kz^2\Psi = \varepsilon\Psi$$

Use Ψ_0 in equation (6) to show that $\varepsilon_o = \frac{1}{2}\hbar\sqrt{\frac{k}{\mu}}$.

Model 2: A Diatomic Molecule.

One model for the vibrations of molecules considers the chemical bond to be similar to a spring attached to two masses. The two masses move back and forth as the spring compresses and expands. This model is very similar to Model 1, the harmonic oscillator. The reduced mass of the two masses is equivalent to the mass in Model 1, and the equilibrium internuclear distance, r_e, is the equivalent of the rest position, $z = 0$, in Model 1.

force constant of bond = k

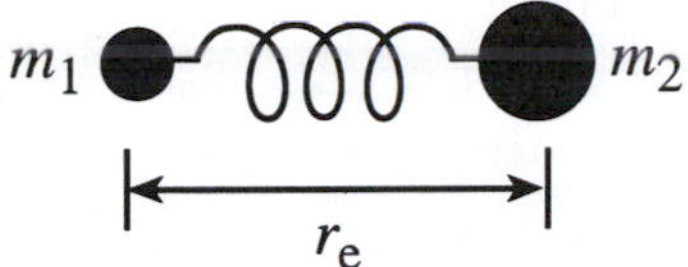

r_e = equilibrium internuclear distance

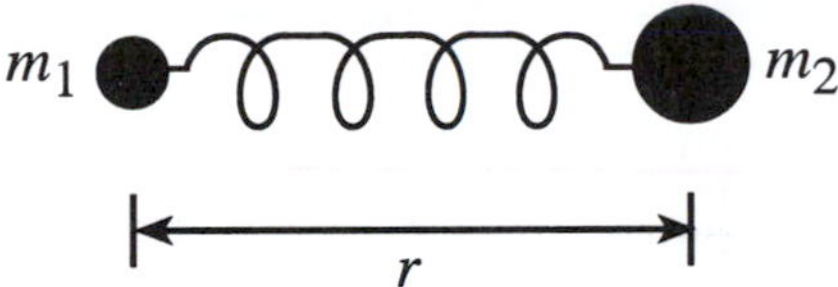

the displacement from the equilibrium internuclear distance = $r - r_e$

$$\mu = \frac{m_1 m_2}{m_1 + m_2} = \text{reduced mass}$$

$$a = \frac{2\pi}{h}\sqrt{k\mu}$$

$$\Psi_0 = \left(\frac{a}{\pi}\right)^{1/4} e^{\frac{-a(r-r_e)^2}{2}} \qquad (8a)$$

$$\Psi_1 = \left(\frac{4a^3}{\pi}\right)^{1/4} (r-r_e)\, e^{\frac{-a(r-r_e)^2}{2}} \qquad (9a)$$

$$\varepsilon_\upsilon = (\upsilon + \tfrac{1}{2})\ \hbar \sqrt{\frac{k}{\mu}} \qquad \upsilon = 0, 1, 2, \ldots \qquad (10)$$

Critical Thinking Questions

4. Suppose a diatomic molecule is in the vibrational state $\upsilon = 0$. Is the vibrational energy also equal to zero? Explain.

5. The vibrational energy of a molecule in the $\upsilon = 0$ state is often referred to as the zero point energy. Is it possible for a diatomic molecule to have less vibrational energy than the zero point energy?

Information

Figure 1. A sketch of the first few energy levels and squares of wave functions for the harmonic oscillator model of a diatomic molecule.

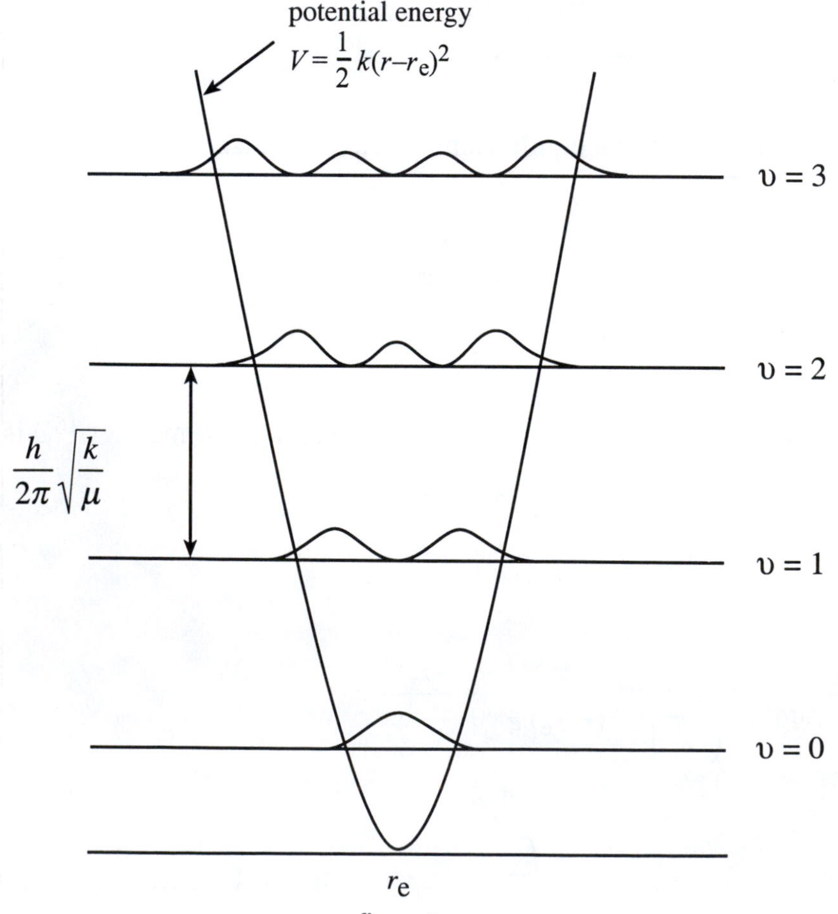

Critical Thinking Questions

6. What is the qualitative relationship between the number of nodes of the wave function and the vibrational energy level?

7. What is the average value of r when the molecule is in a vibrational state described by $\upsilon = 0$? (A numerical value is not being asked for here.)

8. What is the average value of r when the molecule is in a vibrational state described by $\upsilon = 1$?

9. Show that the separation between (successive) energy levels is $\hbar\sqrt{\frac{k}{\mu}}$.

Exercises

3. The molecular mass of ^{12}C is exactly 12 g/mole. The molecular mass of ^{16}O is 15.995 g/mole. Calculate the reduced mass of a $^{12}C^{16}O$ molecule.

4. The force constant for CO is 1860 N m^{-1}. Determine the value of the zero point energy of CO in joules/molecule and kJ/mole.

Model 3: The Relationship Between Bond Energy and Force Constant.

Molecule	Fundamental Transition[a], $\varepsilon_1 - \varepsilon_o$ (cm^{-1})	Force Constant[b] (N m^{-1})	Bond Enthalpy[c], 25°C (kJ/mole)
H_2	4159.5	514	436
D_2	2990.3	531	443

[a]Electromagnetic radiation that increases the vibrational energy of a molecule is usually in the infrared region. Absorption peaks are typically reported with units cm^{-1}, called wavenumbers, (which is proportional to energy). The fundamental transition is from υ = 0 to υ = 1.

[b]The force constants were calculated from $\Delta\varepsilon = \hbar\sqrt{\frac{k}{\mu}}$.

[c]The bond enthalpy at 25°C is ΔH° for the reaction RX(g) → R(g) + X(g).

Critical Thinking Questions

10. Why are the bond enthalpies of H_2 and D_2 almost equal?

11. Note that the values of the force constants for H_2 and D_2 are almost equal. Why are the values of the fundamental transitions (υ = 0 to υ = 1) so different?

12. Based on this analysis, which is the better indicator of bond strength, the fundamental transition energy or the force constant?

Exercises

5. Examine the figure below:

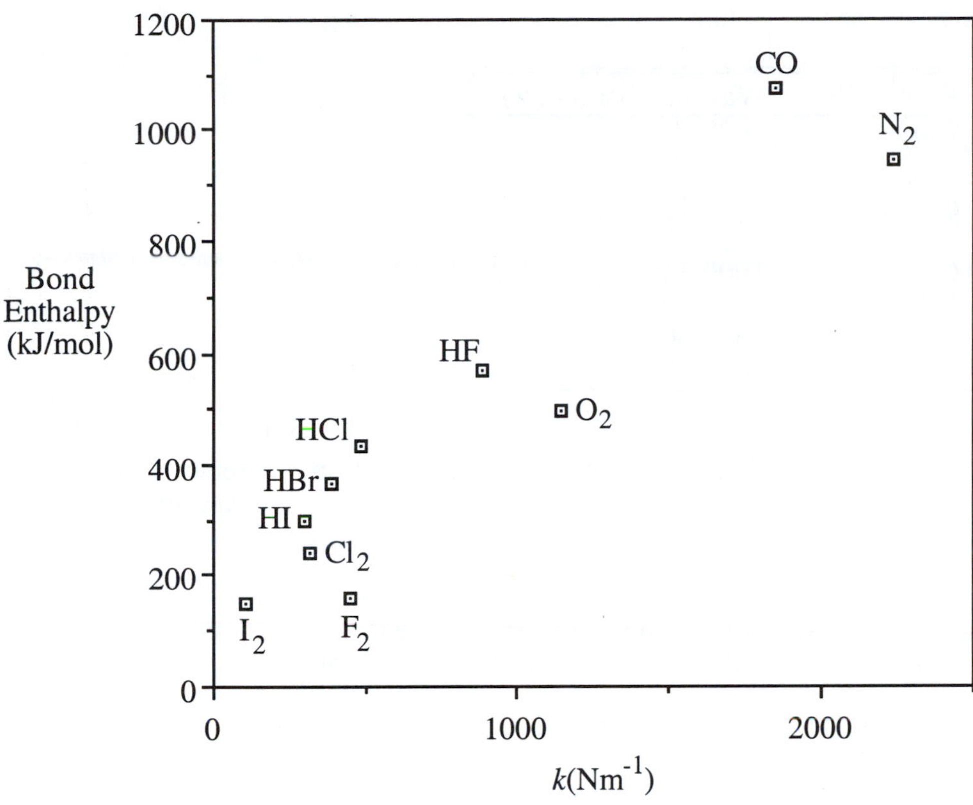

a) Explain the relative bond enthalpies of N_2, O_2 and Cl_2 based on their respective structures. [Hint: Consider the Lewis structures for these species.]

b) Explain the relative bond enthalpies for the series HF, HCl, HBr, HI based on their respective structures.

c) Predict which carbon-carbon bond will have the greater force constant: the bond in ethane or the bond in ethene. Explain your reasoning.

d) Predict which species will have the greater force constant: $Li_2(g)$ or $Na_2(g)$. Explain your reasoning.

e) Which molecule has the shorter bond, HF or HCl? Why?

f) Why is the bond enthalpy of HF greater than the bond enthalpy of HCl?

g) Which molecule has the greater force constant, HF or HCl?

h) Which molecule has the greater force constant, Li_2 or Na_2?

6. In general, which has the largest value for the force constant, a single bond, a double bond, or a triple bond? Explain.

7. Use the Table below and determine if the energies involved are consistent with your answer to Exercise 6. Explain.

Bond Type	Wavenumber (cm^{-1})
C=C	1620–1680
C≡C	2100–2230
C—O	1000-1300
C=O	1630-1800
C—F	1000-1400
C—Cl	600-800
C—Br	500-650
C—I	500-600

8. Explain how the relative bond strengths can be inferred from the relative energies of the C=O and C=C stretches in an IR spectrum. Clearly state any assumptions that you make.

9. a) Based on bond order and bond length, arrange the following bonds in order of increasing bond strength: C–F; C–Cl; C–Br; C–I. Explain.

   b) Is your answer to part a) consistent with the wavenumber values in the Table in Exercise 7? Explain.

10. a) Based on the atomic masses involved, arrange the following bonds in order of increasing reduced mass: C–F; C–Cl; C–Br; C–I. Explain.

    b) Is your answer to part a) consistent with the wavenumber values in the Table in Exercise 7? Explain.

Problems

1. The force constant of HF is 880 N m^{-1}. At what wave number is the fundamental vibrational absorption expected? Where would the corresponding absorption of DF be expected? [Appropriate MWs (g/mole) are: H-1, 1.0078; H-2, 2.014; F-19, 18.998]

2. Predict the fundamental vibrational transition energy of HD as precisely as you can. Explain your analysis clearly and state any assumptions that you make.

ChemActivity **6**

Molecular Rotation

(How fast do molecules rotate?)

The quantum mechanical solution to the harmonic oscillator problem (vibrational motion of diatomic molecule) has produced several important concepts:

- **Quantization** of energy—not all energies are **allowed**.
- A set of **wave functions,** Ψ_i, to describe the particle—each wave function is characterized by a set of **quantum numbers**. The number of quantum numbers used to describe the particle is equal to the number of dimensions.
- More than one **energy state** can exist at a particular **energy level**. Each energy state has a unique set of quantum numbers. The number of energy states that have the same energy is the **degeneracy** of the energy level.
- A set of **probability functions**, Ψ_i^2 (or $\Psi_i^*\Psi_i$), to describe the particle—the particle described by Ψ_i is not uniformly distributed in the space that the particle can occupy.
- The greater the number of **nodes** for the wave function that describes the particle, the higher the energy of the particle.
- The set of wave functions that describe the particle are **orthogonal**.
- Generally, the set of wave functions that describe the particle are **normalized**.

Now, we attempt another three-dimensional problem—the rotational motion of a diatomic molecule.

Model 1: A Particle-on-a-Sphere.

One model for the rotations of molecules considers the chemical bond to be a fixed rod connected to two masses. The masses rotate around the center of mass. This *dumbbell* type arrangement is often referred to as the **rigid rotor**. In its simplest manifestation, the model uses one mass constrained to a fixed distance from the center of a coordinate system. The particle is, therefore, constrained to move on the surface of a sphere.

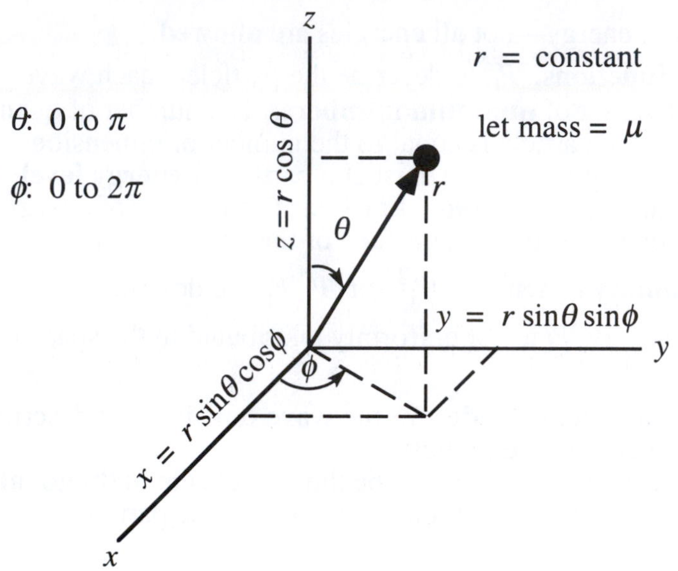

$$U = \frac{1}{2}\,\mu\, v^2 = \frac{1}{2}\,\mu\, r^2 \left(\frac{v}{r}\right)^2 = \frac{1}{2}\, I\omega^2 = \frac{(I\omega)^2}{2I} = \frac{L^2}{2I} \qquad (1)$$

where I is the moment of inertia, $I = \mu r^2$, ω is the angular velocity, $\omega = v/r$, and L is the angular momentum, $L = I\omega$.

$$\hat{H} = \frac{1}{2I}\,\hat{L}^2 \qquad (2)$$

$$\hat{H} = \frac{1}{2I}\,(-\hbar^2)\left\{\frac{1}{\sin\theta}\left[\frac{\partial}{\partial\theta}(\sin\theta)\frac{\partial}{\partial\theta}\right] + \frac{1}{\sin^2\theta}\frac{\partial^2}{\partial\phi^2}\right\} \qquad (3)$$

$$\hat{H}\,\Psi = -\frac{\hbar^2}{2I}\left\{\frac{1}{\sin\theta}\left[\frac{\partial}{\partial\theta}(\sin\theta)\frac{\partial}{\partial\theta}\right] + \frac{1}{\sin^2\theta}\frac{\partial^2}{\partial\phi^2}\right\}\Psi = \varepsilon\Psi \qquad (4)$$

The solutions to equation (4), Ψ_{Jm}, depend on both θ and ϕ, and can be written as a product of two functions, one in θ and one in ϕ.

$$\Psi_{Jm} = \Theta_{Jm}\,\Phi_m \qquad J = 0, 1, 2, \ldots \qquad m = 0, \pm 1, \pm 2, \pm 3, \ldots, \pm J \tag{5}$$

$$\Phi_m = \sqrt{\frac{1}{2\pi}}\; e^{im\phi} \tag{6}$$

Θ_{Jm} = normalized associated Legendre polynomials (see Appendix, Table A.4)

$$\Theta_{Jm} = \mathrm{f}(J, m, \theta) \qquad J = 0, 1, 2, \ldots \qquad m = 0, \pm 1, \pm 2, \pm 3, \ldots, \pm J \tag{7}$$

The first two normalized associated Legendre polynomials are:

$$\Theta_{00} = \frac{\sqrt{2}}{2} \qquad\qquad \Theta_{10} = \frac{\sqrt{6}}{2}\cos(\theta)$$

The rotational energies, ε_J, are given by:

$$\varepsilon_J = \frac{\hbar^2}{2I}\,J(J+1) \qquad\qquad I = \mu r^2 \tag{8}$$

Critical Thinking Questions

1. What is the expression for Φ_0 ?

2. What is the expression for Θ_{00} ?

3. What is the expression for Ψ_{00} ? Ψ_{00}^2 ?

4. a) Explain in physical terms the meaning of Ψ_{00}^2.

b) Based on your answer to CTQ 3, how does the function Ψ_{00} depend on θ? depend on ϕ?

Information

In this coordinate system, the differential element for ϕ is $d\phi$; the differential element for θ is $\sin\theta\, d\theta$. Therefore, the orthogonality and normalization equations are:

$$\int_{\phi=0}^{2\pi}\int_{\theta=0}^{\pi} \Psi_{Jm}^{*}\, \Psi_{Jm} \sin\theta\, d\theta\, d\phi = 1 \quad \text{and} \quad \int_{\phi=0}^{2\pi}\int_{\theta=0}^{\pi} \Psi_{Jm}^{*}\, \Psi_{J'm'} \sin\theta\, d\theta\, d\phi = 0$$

Critical Thinking Questions

5. Show that Ψ_{00} is normalized.

6. According to equation (8), what happens to the rotational energy of the particle if the value of J is increased (holding r and μ constant)?

7. According to equation (8), what happens to the rotational energy of the particle if the mass, μ, is replaced by a heavier mass (holding J and r constant)?

8. According to equation (8), what happens to the rotational energy of the particle if the value of r is increased (hold J and μ constant)?

9. a) If $J = 0$ what values of m are possible?

   b) If $J = 1$ what values of m are possible?

   c) If $J = 2$ what values of m are possible?

10. Derive an expression that gives the number of values of m for a given value of J.

11. Does the energy of a state J,m depend on the value of J? on the value of m?

12. What is the degeneracy of the $J = 3$ level for a rigid rotor?

Exercises

1. Given equation (1), show that equation (3) is correct. [Hint: see CA 2 on Quantum Mechanics.]

2. For Ψ_{00}, show that equation (4) yields the value of the energy as given by equation (8).

3. For Ψ_{10}, show that equation (4) yields the value of the energy as given by equation (8).

4. How many nodes are there in Ψ_{00}? Ψ_{10}? Comment on the relationship between the number of nodes in the rotational wavefunction and the corresponding rotational energy.

5. What is the degeneracy of the $J = 19$ level for a rigid rotor?

Model 2: A Rotating Molecule (rigid).

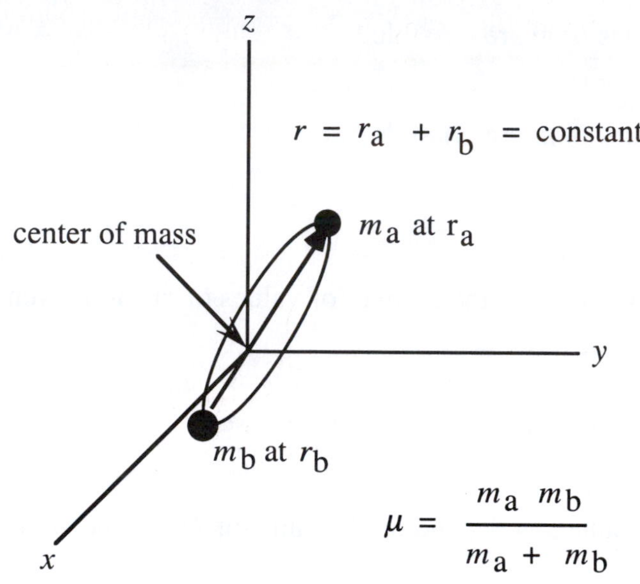

$$\mu = \frac{m_a m_b}{m_a + m_b}$$

Θ = normalized associated Legendre polynomials

$$\Theta_{Jm} = f(J, m, \theta) \quad J = 0, 1, 2, \ldots \quad m = 0, \pm 1, \pm 2, \pm 3, \ldots, \pm J \quad (9)$$

$$\Phi_m = \sqrt{\frac{1}{2\pi}}\, e^{im\phi} \quad (10)$$

$$\varepsilon_J = \frac{\hbar^2}{2I} J(J+1) \qquad I = \mu r^2 \quad (11)$$

The solution is the same as Model 1. Model 1 is a simplified version of a rotating diatomic molecule. The two masses separated by a distance r and rotating about the center of mass in Model 2 are replaced by one equivalent mass, the reduced mass, rotating at a distance r in Model 1.

Critical Thinking Questions

13. If $m_a = m_b$ (homonuclear diatomic), what is the value of μ?

14. The rigid rotor model does not vibrate. Does a real molecule vibrate as it rotates? Comment on the accuracy of this model for a real molecule.

15. Which molecule has the greater internuclear distance (r), H_2 or I_2?

16. Which has the greater reduced mass, H_2 or I_2?

17. Which molecule has the greater rotational energy in the $J = 5$ state, H_2 or I_2?

18. What is the rotational energy of a diatomic molecule described by $J = 0$? What does this imply about the molecular rotation of the molecule?

Exercises

6. The internuclear distance of CO is 113.1 pm. The molecular mass of ^{12}C is exactly 12 g/mole. The molecular mass of ^{16}O is 15.995 g/mole. Determine the value of rotational energy of a $^{12}C^{16}O$ molecule described by $J = 0$. By $J = 1$.

7. Which has more rotational energy in the $J = 1$ state, F_2 or Cl_2?

8. Which has more rotational energy in the $J = 1$ state, H_2 or O_2?

9. If a molecule has rotational energy of 6.67×10^{-23} J and a moment of inertia of 1.00×10^{-45} kg m^2, what is the value of its rotational quantum number?

Problems

The transformation from Cartesian coordinates to polar coordinates is not trivial. As an example of how it is done, consider the transformation from Cartesian coordinates to plane polar coordinates (see below) and problems 1 and 2.

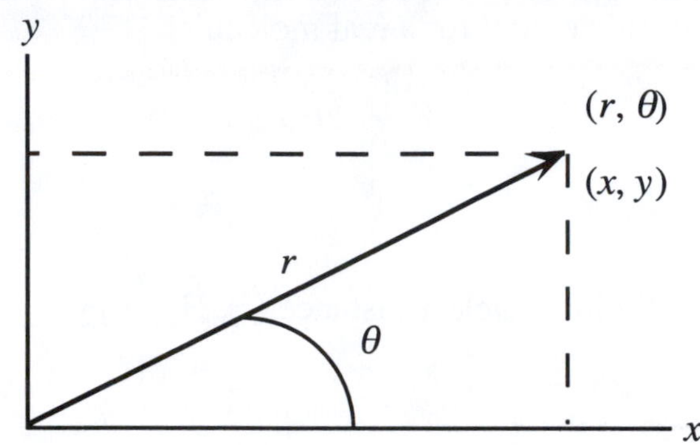

$x = r \cos \theta \qquad y = r \sin \theta \qquad r = (x^2 + y^2)^{1/2}$

1. Recall that if f(x,y) depends on the Cartesian coordinates x and y, then

$$df = \left(\frac{\partial f}{\partial x}\right)_y dx + \left(\frac{\partial f}{\partial y}\right)_x dy \ .$$

Write a similar expression for df if $f(r,\theta)$ depends on the plane polar coordinates r and θ.

2. Use the information in Problem 1 to show that:

$$\left(\frac{\partial f}{\partial x}\right)_y = \left(\frac{\partial f}{\partial r}\right)_\theta \left(\frac{\partial r}{\partial x}\right)_y + \left(\frac{\partial f}{\partial \theta}\right)_r \left(\frac{\partial \theta}{\partial x}\right)_y \qquad \text{Equation A}$$

3. Let a particle be constrained to a circle—called a particle-on-a-ring problem. Here, r is a constant. Use Equation A and the equations from Problem 1: $x = r \cos \theta$ and $y = r \sin \theta$. Show that:

$$\left(\frac{\partial f}{\partial x}\right)_y = \left(\frac{\partial f}{\partial \theta}\right)_r \left(\frac{\partial \theta}{\partial x}\right)_y = \frac{-\sin \theta}{r} \left(\frac{\partial f}{\partial \theta}\right)_r$$

ChemActivity 7

The Hydrogen Atom (I)

(One proton and one electron, is life easy or what?)

Now that we have examined translational, rotational, and vibrational energies, of molecules, we turn our attention to a subject of particular importance to chemists—electronic energies of molecules. We begin with the simplest of all atoms, the hydrogen atom.

Recall that the **electronic energy** is the sum of the following:

- The coulombic interaction of all of the electrons in the molecule with all of the nuclei in the molecule (all of these are attractive forces and negative energies).
- The coulombic interaction of all of the electrons in the molecule with each other (all of these are repulsive forces and positive energies).
- The coulombic interaction of all of the nuclei in the molecule with each other (all of these are repulsive forces and are positive energies).
- The magnetic interaction of all of the particles (electrons and nuclei) in the molecule.
- The kinetic energy of the electrons.

The motion of the center of mass of the molecule, the rotation of the nuclei around the center of mass, and the motion of the nuclei with respect to each other are not part of the electronic energy.

Information

A neutral hydrogen atom consists of one electron bound to a nucleus with a single proton. The quantum mechanical model of the hydrogen atom is very similar to the particle-on-a-sphere model. In this case, however,

- The electron has a negative charge, $-e$.
- The proton has a positive charge, $+e$.
- The proton and the electron move around the center of mass, which is extremely close to the proton. Because the proton is so much more massive than the electron, the reduced mass of the system is essentially equal to the mass of the electron. A simplifying approximation can thus be made, which considers the proton (nucleus) to be fixed (no translational motion) and the electron to move with respect to the proton.
- The distance between the proton and the electron, r, is not constant. The distance r is now a variable.

Model 1: The Hydrogen Atom.

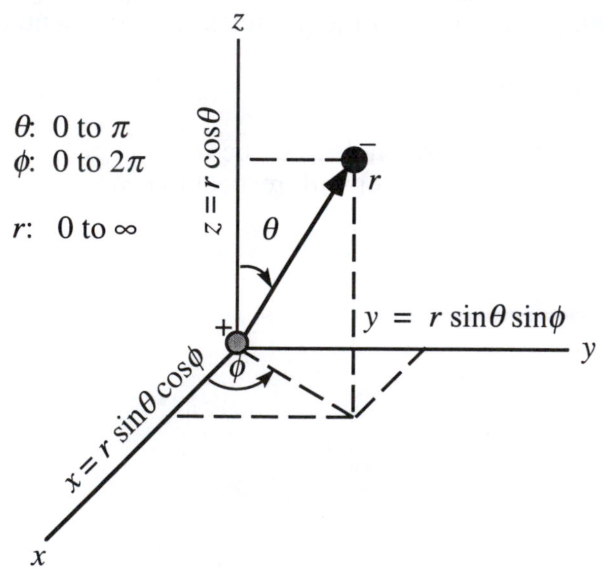

$$U = \frac{1}{2} m_e v^2 + \frac{(-e)(+e)/4\pi\varepsilon_o}{r} = \frac{1}{2} m_e v^2 - \frac{e^2/4\pi\varepsilon_o}{r}$$

where ε_o is the vacuum permittivity. Thus,

$$H = \frac{p^2}{2m} - \frac{e^2/4\pi\varepsilon_o}{r} \qquad (1)$$

It is convenient to solve this problem in spherical coordinates because the system has a natural center of symmetry (the electron revolves around proton at the origin). See Table 1 of ChemActivity 2 on Quantum Mechanics.

$$\hat{H} = -\frac{\hbar^2}{2m}\left\{\frac{1}{r^2}\frac{\partial}{\partial r}\left(r^2\frac{\partial}{\partial r}\right)+\frac{1}{\sin\theta}\left[\frac{\partial}{\partial\theta}(\sin\theta)\frac{\partial}{\partial\theta}\right]+\frac{1}{\sin^2\theta}\frac{\partial^2}{\partial\phi^2}\right\} - \frac{e^2/4\pi\varepsilon_o}{r} \quad (2)$$

$$\hat{H}\,\Psi = -\frac{\hbar^2}{2m}\left\{\frac{1}{r^2}\frac{\partial}{\partial r}\left(r^2\frac{\partial}{\partial r}\right)+\frac{1}{\sin\theta}\left[\frac{\partial}{\partial\theta}(\sin\theta)\frac{\partial}{\partial\theta}\right]+\frac{1}{\sin^2\theta}\frac{\partial^2}{\partial\phi^2}\right\}\Psi - \frac{e^2/4\pi\varepsilon_o}{r}\Psi = \varepsilon\Psi \quad (3)$$

The solutions to equation (3) are a product of three functions: $R_{n\ell}$, a function of r; $\Theta_{\ell m}$, a function of θ; Φ_m, a function of ϕ.

$$\Psi_{n\ell m} = R_{n\ell}\Theta_{\ell m}\Phi_m \quad (4)$$

$R_{n\ell}$ = the Radial function = $f(r,n,\ell)$ (see Appendix)

$n = 1, 2, 3, \ldots$

$\ell = 0, 1, 2, 3, \ldots, n-1$

The first few $R_{n\ell}$ are given in Table 1.

Table 1. Several Radial Functions for the Hydrogen Atom.

$$R_{10} = 2\left(\frac{1}{a_o}\right)^{3/2} e^{-r/a_o}$$

$$R_{20} = \left(\frac{1}{2a_o}\right)^{3/2}\left(2-\frac{r}{a_o}\right) e^{-r/2a_o}$$

$$R_{21} = \frac{1}{\sqrt{3}}\left(\frac{1}{2a_o}\right)^{3/2}\frac{r}{a_o}e^{-r/2a_o}$$

where a_0 is the Bohr radius, $a_o = \dfrac{\hbar^2}{m_e\,(e^2/4\pi\varepsilon_o)} = 52.918$ pm

Φ_m and $\Theta_{\ell m}$ are the same functions as found for the particle-on-a-sphere (see ChemActivity 6 on Molecular Rotation). The quantum number J used in the particle-on-a-sphere solutions, is replaced by the quantum number ℓ for the hydrogen atom. Typically, the product of these two functions is taken: $Y_{\ell m} = \Theta_{\ell m}\Phi_m$. The functions $Y_{\ell m}$ are known as the normalized spherical harmonics and are often called the angular functions.

$Y_{\ell m} = \Theta_{\ell m}\Phi_m \quad \ell = 0, 1, 2, 3, \ldots, n-1 \quad m = 0, \pm1, \pm2, \pm3, \ldots, \pm\ell$

$$\mathcal{E}_n = -\frac{\hbar^2}{2m_e a_0^2}\frac{1}{n^2} = -\frac{1}{n^2}\,2.179 \times 10^{-18}\text{ J} \qquad n = 1, 2, 3, \ldots \tag{5}$$

Critical Thinking Questions

1. The electronic energies for the hydrogen atom are negative. Is the physical reason for this embodied in the kinetic energy or in the potential energy (Coulomb's law)?

2. Which is the more negative energy level, $\mathcal{E}_1$ or $\mathcal{E}_4$?

3. Which state of the hydrogen atom has the lower energy, $n = 1$ or $n = 4$?

4. Write the expression for the reduced mass, μ, for the hydrogen atom. Recall that m_p/m_e is approximately 1836. Show that the reduced mass is very close to m_e.

5. How does the energy, $\mathcal{E}$, depend on the quantum numbers ℓ and/or m?

6. For R_{10}, for what values of r is

   a) $R_{10} = 0$?

   b) $R_{10} > 0$?

   c) $R_{10} < 0$?

7. For R_{20}, for what values of r is

   a) $R_{20} = 0$?

   b) $R_{20} > 0$?

   c) $R_{20} < 0$?

8. For R_{21}, for what values of r is

   a) $R_{21} = 0$?

   b) $R_{21} > 0$?

   c) $R_{21} < 0$?

9. Using grammatically correct English sentences, describe the radial functions, making particular note of the values of $r = 0$ and as $r \longrightarrow \infty$, and the location of any radial nodes. A response for R_{10} has been provided as an example.

   R_{10}: This function is positive at r = 0, has no radial nodes, and approaches 0 as $r \longrightarrow \infty$. It is positive at all values of r.

   a) R_{20}:

   b) R_{21}:

10. A node at a specific non-zero value of r is often referred to as a **radial node**. Which geometric term best describes a radial node in space: point, line, plane, spherical shell, cubical shell, conical shell?

Exercise

1. Given equation (1), show that the Hamiltonian operator in equation (2) is correct. [Hint: see CA 2 on Quantum Mechanics.]

Information

A one-electron **wavefunction** is a mathematical description of an electron associated with an atom (or ion) that has only one electron.

Model 2: A Hydrogen Atom Wavefunction, Ψ_{100}.

$$\Psi_{100} = \frac{1}{\pi^{1/2}}\left(\frac{1}{a_o}\right)^{3/2} e^{-r/a_o}$$

a_o is the Bohr radius, 52.918 pm
r is the radial coordinate of the electron
$r = (x^2 + y^2 + z^2)^{1/2}$

Critical Thinking Questions

11. Using the expressions for $R_{n\ell}$, $\Theta_{\ell m}$, and Φ_m show that the expression for Ψ_{100} for the hydrogen atom in Model 2 is correct.

12. For the hydrogen atom, determine the numerical value (including units) of Ψ_{100} at each of the following points:

    a) $x = 0, y = 0, z = 0$

b) $x = a_o, y = 0, z = 0$

c) $x = -a_o, y = 0, z = 0$

d) $x = 0, y = 0, z = a_o$

e) $x = 0, y = a_o, z = 0$

f) $x = \left(\frac{a_o^2}{3}\right)^{1/2}, \ y = -\left(\frac{a_o^2}{2}\right)^{1/2}, \ z = \left(\frac{a_o^2}{6}\right)^{1/2}$

13. For the hydrogen atom, what is the numerical value of Ψ_{100} at $r = a_o$?

14. For the hydrogen atom, what is the numerical value of Ψ_{100} at $r = \infty$?

15. For the hydrogen atom, make a sketch of Ψ_{100} versus r (0 to ∞) on the axes below. In a sketch, it is important to show where the function is positive, where the function is negative, where the function is zero (a node); actual numerical values are not given except in the case of nodes.

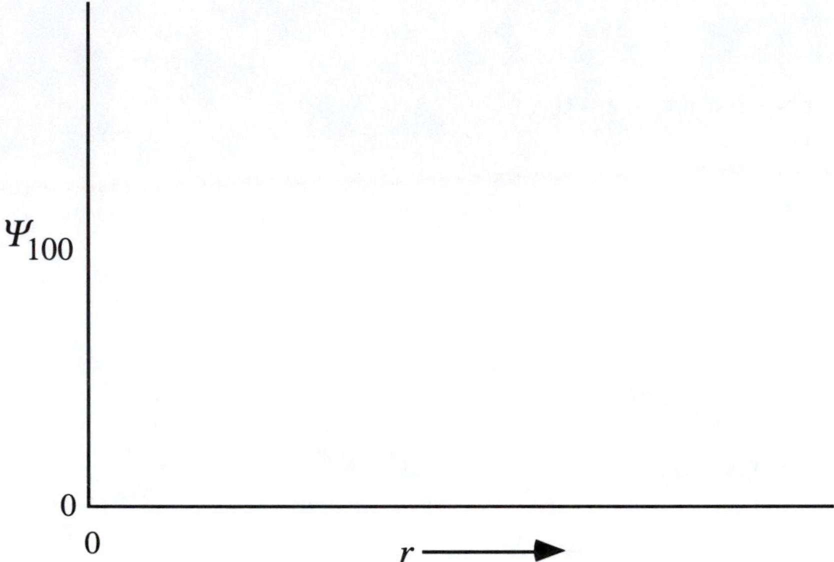

Model 3: Graphical Representation of $\Psi_{100} = \Psi_{1s}$.

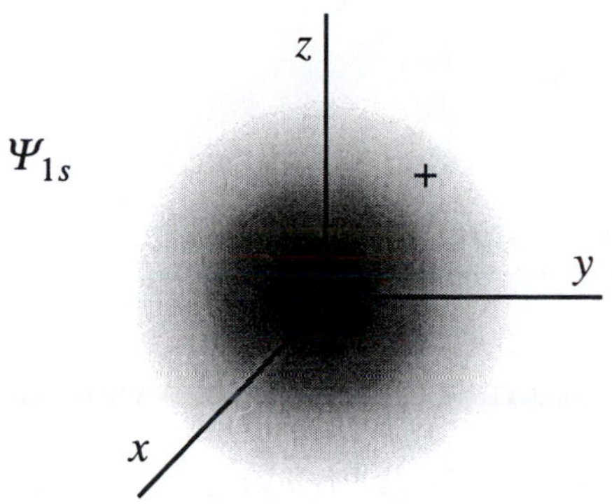

Wavefunctions for which $\ell = 0$ are said to be *s* wavefunctions.

Critical Thinking Questions

16. Based on Model 2 and your previous answers,

    a) Why is a positive sign found in the graphical representation of Ψ_{1s} (Ψ_{100}) in Model 3?

    b) Why is a circle or a sphere used in the graphical representation of Ψ_{1s} (Ψ_{100}) in Model 3?

Model 4: A Hydrogen Atom Wavefunction, Ψ_{210}.

$$\Psi_{210} = \frac{1}{4(2\pi)^{1/2}} \left(\frac{1}{a_o}\right)^{3/2} \frac{z}{a_o} e^{-r/2a_o}$$

where $z = r\cos\theta$

Critical Thinking Questions

17. For the hydrogen atom, determine the numerical value of Ψ_{210} (including units) at

a) $x = 0, y = 0, z = 0$

b) $x = 0, y = 0, z = a_o$

c) $x = 0, y = 0, z = -a_o$

18. For the hydrogen atom, what is the numerical value of Ψ_{210} at $x = 0$, $y = 0$, as z approaches infinity?

19. $\Psi_{210} = 0$ when $z = 0$. That is $\Psi_{210} = 0$ for this wavefunction for all values of x and y when $z = 0$. Which geometric term best describes this node in space: point, line, plane, spherical shell, cubical shell, conical shell?

20. For the hydrogen atom, make a sketch of Ψ_{210} versus z ($-\infty$ to ∞) on the axes below. In a sketch, it is important to show where the function is positive, where the function is negative, where the function is zero (a node); actual numerical values are not given except in the case of nodes.

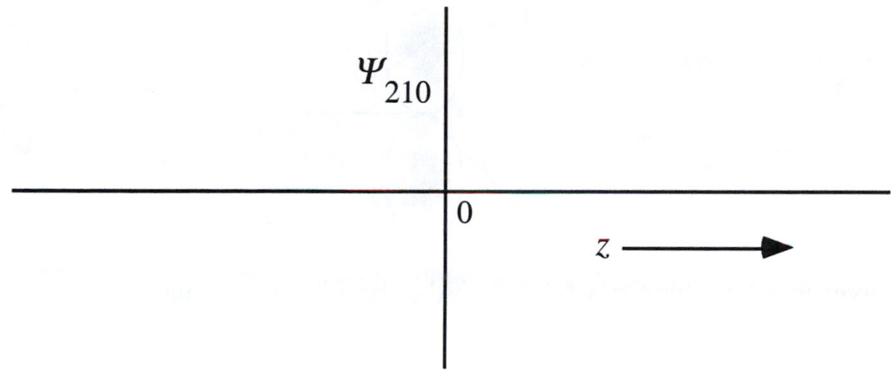

Model 5: Graphical Representation of $\Psi_{210} = \Psi_{2p_z}$

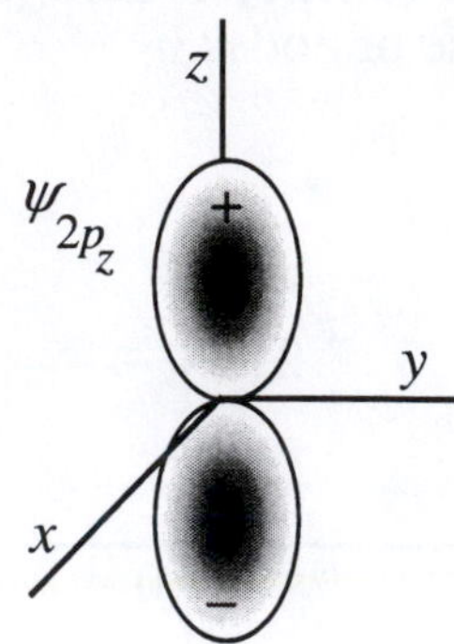

A wavefunction which has $\ell = 1$ is said to be a "p" wavefunction.

Critical Thinking Questions

21. Why is the wavefunction in Model 5 labeled $2p_z$, as opposed to $2p_x$ or $2p_y$?

22. Based on Models 4 and 5 and your answers to previous questions,

    a) Why is a positive sign found in the graphical representation of Ψ_{2p_z} ?

    b) Why is a negative sign found in the graphical representation of Ψ_{2p_z} ?

    c) Why does the graphical representation of Ψ_{2p_z} have the shape of a dumbbell?

Information

The general form for the wavefunctions for one-electron species (H, He^+, Li^{2+}, and so on) includes the atomic number, Z, of the species. For example,

$$\Psi_{1s} = \frac{1}{\pi^{1/2}}\left(\frac{Z}{a_o}\right)^{3/2} e^{-Zr/a_o}$$

$$\Psi_{2pz} = \frac{1}{4(2\pi)^{1/2}}\left(\frac{Z}{a_o}\right)^{3/2} \frac{zZ}{a_o} e^{-Zr/2a_o}$$

The general functional forms for one-electron wavefunctions can be found in the Appendix, Table A.8.

The energy of a one-electron wavefunction is given by

$$\varepsilon_n = -\frac{\hbar^2}{2m_e a_o^2}\frac{Z^2}{h^2}$$

Exercises

2. Every wavefunction for a one-electron atom is a product of two functions (or three, depending on how you count them)—a radial function and an angular function. Show that Ψ_{2p_z} is a product of R_{21} and Y_{10}.

3. Calculate the value of Ψ_{1s} for the hydrogen atom, the He^+ ion, and the Li^{2+} ion at $r = 0$, $r = a_o$, and $r = 4\ a_o$. Rationalize the trends in the values that you obtain.

4. $\Psi_{2s} = \frac{1}{4\sqrt{2\pi}}\left(\frac{Z}{a_o}\right)^{3/2}\left(2 - \frac{Zr}{a_o}\right)\exp\left(\frac{-Zr}{2a_o}\right)$. For a hydrogen atom, what is the numerical value of Ψ_{2s} at $r = 0$? At $r = a_o$? At $r = 2a_o$? At $r = 3a_o$? At $r = \infty$?

5. For the hydrogen atom, make a sketch of Ψ_{2s} versus r (0 to ∞). How would the location of any radial nodes change as the nuclear charge is increased (as, for example in He^+)? See the Appendix, Table A.8, to check your answer.

6. A graphical representation of Ψ_{2p_y} is shown below. Can you guess the mathematical formula of Ψ_{2p_y} ?

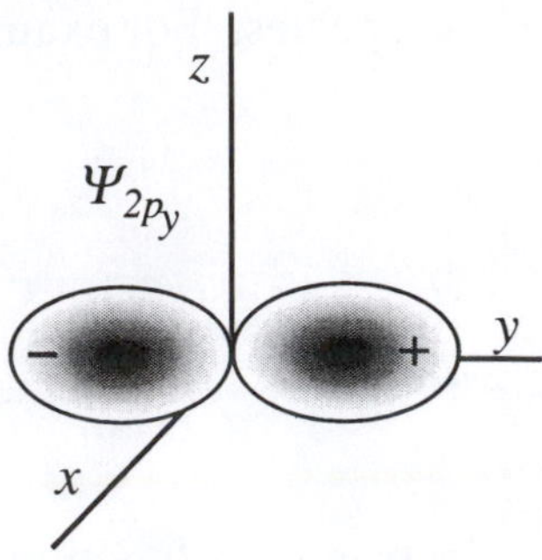

7. For which one-electron atom would you expect that a greater electron density would be found very close to the nucleus, H or He^+? Briefly explain?

8 For which one-electron atom would you expect that a greater electron density would be found very far from the nucleus, H or He^+? Briefly explain?

9. The wavefunction for the $3d_{xz}$ orbital is:

$$\Psi_{3d_{xz}} = = \frac{\sqrt{2}}{81\sqrt{\pi}}\left(\frac{Z}{a_o}\right)^{3/2}\left(\frac{Zr}{a_o}\right)^2 \exp\left(\frac{-Zr}{3a_o}\right) \sin\theta \cos\theta \cos\Phi .$$

A graphical representation of this orbital is shown below.

(a) Explain why there is a node at $r = 0$.
(b) Explain why the function is positive in the region of space where x is positive and z is positive.
(c) Explain why the function is negative in the region of space where x is negative and z is positive.

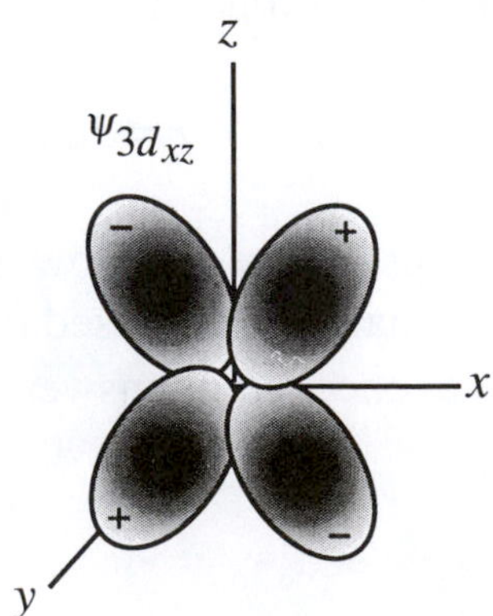

ChemActivity 8

The Hydrogen Atom (II)

(What Else Do We Know About Hydrogen?)

We have seen that the electronic energies of the hydrogen atoms are quantized and that the single electron can occupy regions of space called orbitals. Now, we turn our attention to the **angular momentum** of the single electron in the atom, the **probability** of finding the electron at various locations in space, and the **transition energy** between different energy states.

Model 1: Eigenfunctions and Eigenvalues.

For the hydrogen atom model:

$$\hat{H}\,\Psi_{n\ell m} = -\frac{\hbar^2}{2m_e a_0^2}\frac{1}{n^2}\,\Psi_{n\ell m} \tag{1}$$

$$\hat{L}^2\,\Psi_{n\ell m} = \ell(\ell+1)\,\hbar^2\,\Psi_{n\ell m} \tag{2}$$

$$\hat{L}_z\,\Psi_{n\ell m} = m\hbar\,\Psi_{n\ell m} \tag{3}$$

Note that in all three of these equations, when the operator is applied to the function (in this case, $\Psi_{n\ell m}$), the result is a constant multiplied by the original function. Equations of this type are referred to as **eigenvalue** equations, and the resulting constant is called an **eigenvalue** of the operator. Any *function* for which the equation is valid is an **eigenfunction** of the operator.

If the wavefunction $\Psi_{n\ell m}$ is an eigenfunction of a particular operator, then the state of the system described by the wavefunction has a definite value for the physical property represented by the operator. For example, since $\Psi_{n\ell m}$ is an eigenfunction of the total energy operator $\hat{H}$, we say that the state described by the wavefunction $\Psi_{n\ell m}$ has a definite total energy, and it is given by the eigenvalue: $-\frac{\hbar^2}{2m_e a_0^2}\frac{1}{n^2}$.

If $\Psi_{n\ell m}$ is *not* an eigenfunction of a particular operator (that is, when the operator is applied to $\Psi_{n\ell m}$, the result is *not* a constant multiplied by $\Psi_{n\ell m}$), then the physical property associated with the operator *does not* have a definite value for the state described by $\Psi_{n\ell m}$.[3]

Critical Thinking Questions

1. a) What are the n, ℓ, and m values for Ψ_{100} of the hydrogen atom?

   b) What is the value of the square of the angular momentum, L^2, for Ψ_{100} of the hydrogen atom? Recall that the value of Planck's constant bar, $\hbar$, is 1.055×10^{-34} Js.

[3]Nonetheless, as we have seen previously, an average value for the physical property q, <q> can be obtained.

2. a) What are the n, ℓ, and m values for Ψ_{210} of the hydrogen atom?

b) What is the value of the square of the angular momentum, L^2, for Ψ_{210} of the hydrogen atom?

3. For a given $\Psi_{n\ell m}$ of the hydrogen atom does the square of the angular momentum, L^2, have a definite value? If so, what is it? You may give your answer in terms of $\hbar^2$ such as: $5\hbar^2$ or $n^2\hbar^2$.

4. For a given $\Psi_{n\ell m}$ of the hydrogen atom does the z-component of the angular momentum, L_z, have a definite value? If so, what is it? You may give your answer in terms of $\hbar$ such as: $5\,\hbar$ or $n^2\,\hbar$.

5. What is the value (eigenvalue) of L_z for the wavefunction $\Psi_{2,1,-1}$? You may give your answer in terms of $\hbar$ such as: $5\hbar$ or $n^2\hbar$.

6. What is the value of L^2 for the wavefunction $\Psi_{2,1,-1}$? You may give your answer in terms of $\hbar^2$ such as: $5\,\hbar^2$ or $n^2\hbar^2$.

7. What is the expression for the magnitude of L, $|L|$, for the wavefunction $\Psi_{n\ell m}$?

8. What is the value of the magnitude of L, $|L|$, for the wavefunction $\Psi_{2,1,-1}$? You may give your answer in terms of $\hbar$ such as: $5\hbar$ or $n^2\hbar$.

9. In order for equations (2) and (3) to be reasonable, it must be the case that the z-component of the angular momentum is smaller than or equal to the magnitude of the angular momentum. Expressed algebraically, $L_z \leq |L|$.

   a) Use a vector diagram (for the angular momentum and for the z-component of the angular momentum) to explain why this must be true in general (without reference to equations (2) and (3)).

   b) Show that this *is* true using equations (2) and (3) and the restriction on n, ℓ, and m:

   $n = 0, 1, 2, \ldots$
   $\ell = 0, 1, 2, \ldots, n-1$
   $m = 0, \pm 1, \pm 2, \ldots, \pm \ell$

Exercises

1. An electron of a hydrogen atom is described by $n = 4$, $\ell = 3$, and $m = -2$, $\Psi_{4,3,-2}$. a) What is the energy, in joules? b) What is the value of the magnitude of the angular momentum, L? c) What is the value of the z component of the angular momentum, L_z?

2. Deduce the value of the magnitude of the total angular momentum, L, for any p orbital. For any d orbital.

3. Show that equations (2) and (3) are correct for Ψ_{100}.

4. Show, explicitly, that $\hat{L}_z \, \Psi_{n\ell m} = m\hbar \, \Psi_{n\ell m}$.

5. Show that equation (2) is correct for Ψ_{210}.

6. What is the value of L^2 for the wavefunction $\Psi_{4,3,1}$? You may give your answer in terms of $\hbar^2$.

7. The operator for the position of the electron is $\hat{\boldsymbol{r}} = \boldsymbol{r}$. Is Ψ_{100} an eigenfunction of $\hat{\boldsymbol{r}}$? If so, what is the eigenvalue? If not, what does this suggest about the position of the electron? [Hint: Here, $\boldsymbol{r}$ is the vector indicating the position in space of the electron: (r, θ, ϕ) or (x, y, z). Because Ψ_{100} depends only on the magnitude of $\boldsymbol{r}$, the operator may be thought of as "multiply by r".]

Model 2: The Position of the Electron in the Hydrogen Atom.

We have seen that the wavefunction Ψ_{100} is not an eigenfunction of the position operator $\hat{r}$. This result can be shown to be true for all of the hydrogen atom wavefunctions $\Psi_{n\ell m}$. Rather, the position of the electron in $\Psi_{n\ell m}$ is interpreted in a probabilistic manner. The probability of finding the electron between r_i and r_j, θ_i and θ_j, and ϕ_i and ϕ_j is given by:

$$P_{r,\theta,\phi} = \int_{r_i}^{r_j} \int_{\phi_i}^{\phi_j} \int_{\theta_i}^{\theta_j} \Psi^2 \, d\tau = \int_{r_i}^{r_j} \int_{\phi_i}^{\phi_j} \int_{\theta_i}^{\theta_j} R^2 \Theta^2 \Phi^2 \, d\tau \tag{4}$$

where $d\tau$ is the volume element in spherical coordinates,

$$d\tau = r^2 \sin\theta \; d\theta \, d\phi \, dr \tag{5}$$

$$P_{r,\theta,\phi} = \int_{r_i}^{r_j} \int_{\phi_i}^{\phi_j} \int_{\theta_i}^{\theta_j} R^2 \Theta^2 \Phi^2 \; r^2 \sin\theta \; d\theta \, d\phi \, dr \tag{6}$$

Integration over all values of θ (0 to π) and all values of ϕ (0 to 2π) yields the probability of finding the electron between r_i and r_j:

$$P_r = \int_{r_i}^{r_j} r^2 R^2 \, dr \tag{7}$$

The function $r^2 R^2$ is called the **radial distribution function**. This function is defined as the probability per unit value of r of finding the electron at a distance r from the nucleus.. That is,

$r^2 R^2 \, dr$ = The probability that the particle lies at a distance from the nucleus between r and $r + dr$

Figure 1. The radial distribution function for the 1*s*, 2*s* and 2*p* orbitals.

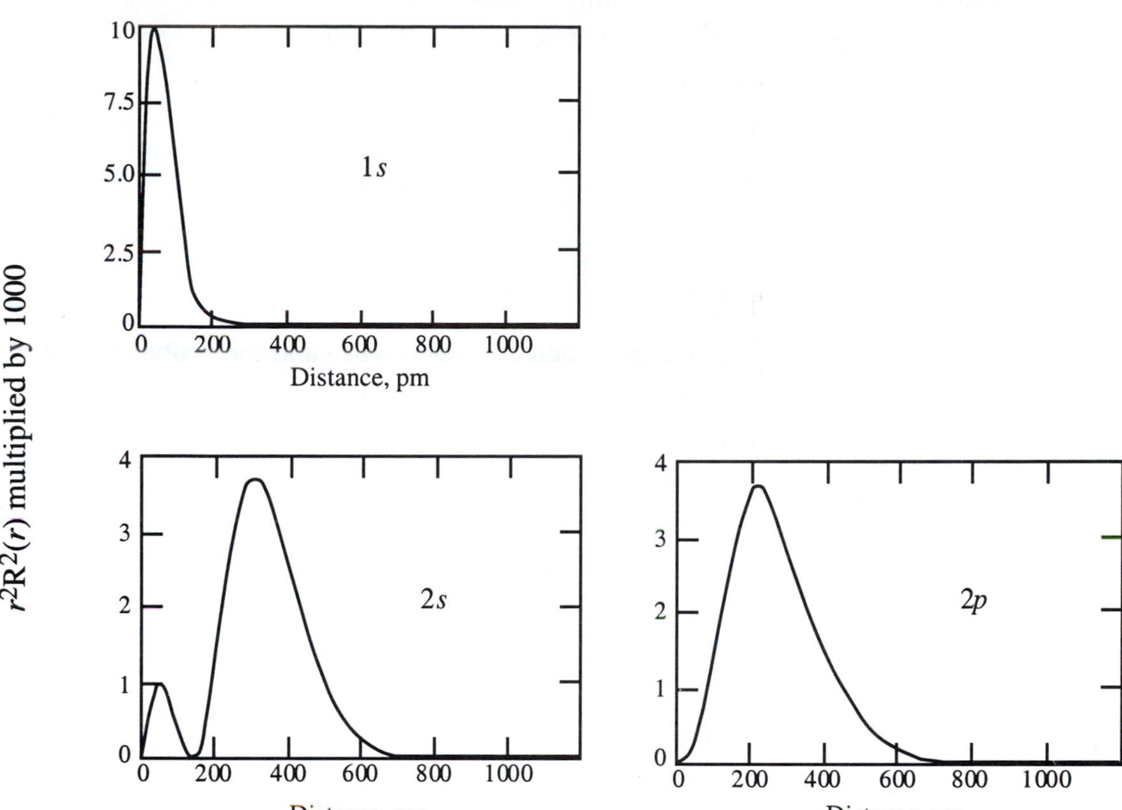

This function can be used to determine the most probable distance of the electron from the nucleus. For Ψ_{100}, the most probable distance of the electron from the nucleus, r_{mp}, can be shown to be:

$$r_{mp} = a_o \qquad (8)$$

Critical Thinking Questions

10. Use equation (7) to write an expression that, when evaluated, will yield the value of the probability of finding the electron between $r = a_o$ and $r = 2a_o$.

11. Use equation (7) to write an expression which shows that the probability of finding the electron *somewhere* in the universe is 1.

12. Describe, using grammatically correct English sentences, how equation (8) can be obtained from the radial distribution function.

Exercises

8. Show how equation (7) is obtained from equation (6).

9. Carefully set up the integral that when evaluated will yield the probability of finding the electron in Ψ_{100} of the hydrogen atom at r values between r = 0 and r = r_{mp}. You may start with equation (7).

10. Recall that

$$\Psi_{2pz} = \frac{1}{4(2\pi)^{1/2}} \left(\frac{1}{a_o}\right)^{3/2} \frac{r \cos \theta}{a_o} \; e^{-r/2a_o} \quad .$$

Carefully <u>set up</u> the integral in terms of r, θ, and ϕ, that will, when evaluated, give the probability of finding an electron in a $2p_z$ orbital at locations where $r \leq a_o$ and x, y, and z are positive—as shown in the figure below.

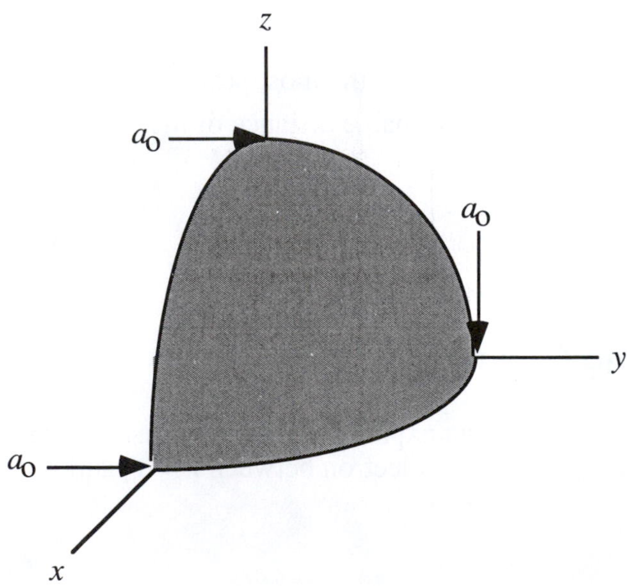

11. Show explicitly how equation (8) can be obtained for a *1s* orbital.

12. Calculate the average value of r, $\langle r \rangle$, for Ψ_{100}. Compare this value to r_{mp} and rationalize the result based on the representation of r^2R^2 for Ψ_{100} in Figure 1.

Model 3: Electronic Transitions of the Hydrogen Atom.

$$\mathcal{E}_n = -\frac{\hbar^2}{2m_e a_0^2}\frac{1}{n^2} = -\frac{1}{n^2}\,2.179 \times 10^{-18}\text{ J} \quad n = 1, 2, 3, \ldots \qquad (9)$$

$\mathcal{E}_1$ is called "the ground state" (the lowest energy state)

Critical Thinking Questions

13. What is the energy, in joules, of

    a) the $n = 1$ state?

    b) the $n = \infty$ state?

14. What is the value of the ionization energy for the ground state of the hydrogen atom (the energy needed to completely remove the electron from the nucleus). Express your answer in joules per atom and MJ per mole.

15. From which state of the hydrogen atom is it harder to remove the electron: $\Psi_{2,1,0}$ or $\Psi_{5,4,2}$?

    a) Use equation (9) to explain.

    b) Use Coulomb's law to explain.

16. If a hydrogen atom electronic transition occurs from $\mathcal{E}_{2,1,0}$ to $\mathcal{E}_{1,0,0}$ is a photon absorbed by the atom or released by the atom? Explain.

Information

The set of emissions from excited hydrogen atoms to the ground state ($n = 1$) is referred to as the "Lyman series" and the set of emissions from excited hydrogen atoms to the $n = 2$ state is referred to as the "Balmer series".

Critical Thinking Question

17. Which series includes higher energy photons, the Lyman or the Balmer series? Explain without calculating the precise energy of any photons.

Exercises

13. Calculate the energy (in joules) of the photon released for the lowest energy transition for the Lyman series and for the Balmer series.

14. Which of the following describe(s) an excited state of hydrogen: $\Psi_{2,1,0}$; $\Psi_{1,0,0}$; $\Psi_{8,1,0}$; $\Psi_{5,4,-2}$?

15. a) Find an electronic transition for a hydrogen atom where a visible photon (400 nm to 800 nm) is released. b) What is the n quantum number for the state of the atom after the photon is released?

16. Calculate the energy (in joules), the frequency, and wavelength (in nm), and the wavenumber (in cm^{-1}) for the first five lines of the Balmer series of the hydrogen atom spectrum. In what range(s) of the electromagnetic spectrum are these lines found?

ChemActivity 9

Multielectron Atoms

(Why does life have to be so complicated?)

Fortunately, there are atoms with more than one electron, **multielectron atoms**. Even hydrogen can have more than one electron—the hydride ion. If there weren't multielectron atoms, we would not exist. The theoretical explanation of the energies and the properties of these atoms has proven to be a formidable task for scientists. We will begin with the simplest multielectron atom—He.

Information

As a shorthand notation for writing more complicated Schrödinger equations, we make the following definition:

$$\nabla^2 = \frac{\partial^2}{\partial x^2} + \frac{\partial^2}{\partial y^2} + \frac{\partial^2}{\partial z^2} \quad (1)$$

The Laplacian operator, ∇^2, is read "del-squared".

Model 1: The Helium Atom.

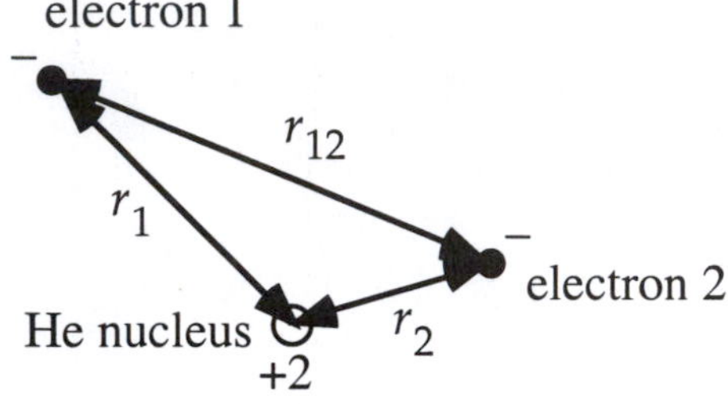

$$U = \frac{1}{2} m_e v_1^2 + \frac{1}{2} m_e v_2^2 + \frac{(2e)(-e)/4\pi\varepsilon_o}{r_1} + \frac{(2e)(-e)/4\pi\varepsilon_o}{r_2} + \frac{(-e)(-e)/4\pi\varepsilon_o}{r_{12}}$$

$$U = \frac{1}{2} m_e v_1^2 + \frac{1}{2} m_e v_2^2 - \frac{2e^2/4\pi\varepsilon_o}{r_1} - \frac{2e^2/4\pi\varepsilon_o}{r_2} + \frac{e^2/4\pi\varepsilon_o}{r_{12}} \quad (2)$$

$$\hat{H}\Psi = \left(\frac{\hbar^2}{2m_e}(\nabla_1^2 + \nabla_2^2) - \frac{Ze^2/4\pi\varepsilon_o}{r_1} - \frac{Ze^2/4\pi\varepsilon_o}{r_2} + \frac{e^2/4\pi\varepsilon_o}{r_{12}}\right)\Psi = \varepsilon\Psi \quad (3)$$

Equation (3) provides a general expression for any two-electron species, for example He, Li^+, Be^{2+}, etc.

Critical Thinking Questions

1. The velocities v_1 and v_2 are labeled separately in equation (2). Why isn't the mass of the electron labeled m_{e1} and m_{e2}?

2. There is a "$2e^2$" in the third and fourth terms of the right-hand-side of equation (2). Why is there no "$2e^2$" in the fifth term?

3. There is a negative sign in front of the third and fourth terms of equation (2). Why? Why is there a positive sign in front of the fifth term?

4. What does the Z represent in equation (3)? What is the value of Z for He? Li^+? Be^{2+}?

Information

Mathematicians, chemists, physicists, and other industrious individuals have tried to find a function which is a solution to equation (3). None has succeeded. The reason for this is that this equation cannot be solved. There are no functions which are eigenfunctions of the Hamiltonian operator in equation (3).

Critical Thinking Question

5. Is this course over?

Information

For all atomic and molecular species with more than one electron, the Schrödinger equation cannot be solved. However, it is possible to generate functions which are *approximate* solutions. The variational method is one of the approximation methods used when the solution to the Schrödinger equation cannot be found, as is usually the case. The variational method is summarized below:

- Write the Hamiltonian in the usual fashion and write $\hat{H}\,\Psi = \varepsilon\Psi$.
- If the Schrödinger equation cannot be solved, make an educated guess—called the trial solution, Ψ_{trial}. (The trial solution must obey the boundary conditions.)
- Determine the average value of ε.

$$\langle\varepsilon\rangle = \frac{\int \Psi^*_{\text{trial}}\hat{H}\Psi_{\text{trial}}d\tau}{\int \Psi^*_{\text{trial}}\Psi_{\text{trial}}d\tau} \tag{4}$$

  It can be shown that average value of the energy will be too high when this equation is used for the average value of the energy. That is, the true energy of the system is always **lower** than the value calculated with equation (4).

- If necessary, make another educated guess for Ψ_{trial} and recalculate the average value for the energy.
- The lower the energy of the trial solution, the closer the trial solution is to the real solution of the Schrödinger equation. (Remember, it is not possible to find a trial solution that yields an energy lower than the actual energy of the system.)

Model 2: Trial Function for a Particle-on-a-Line.

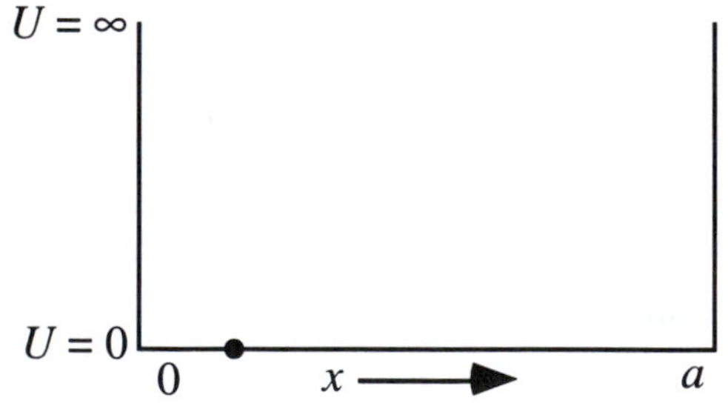

$U = \infty$

$U = 0$

0 x ⟶ a

$$\hat{H}\,\Psi = \frac{-h^2}{8m\pi^2}\frac{d^2\Psi}{dx^2} = \varepsilon\Psi \tag{5}$$

$$\Psi_{\text{trial}} = B\,[x(a-x)] \tag{6}$$

$$\langle\varepsilon\rangle = \frac{10}{\pi^2}\frac{h^2}{8ma^2} \tag{7}$$

Critical Thinking Questions

6. The boundary condition for a particle-on-a-line is that $\Psi = 0$ at $x = 0$ and $\Psi = 0$ at $x =$ a. Does the trial function obey the boundary condition?

7. Compare the average value of the energy, equation (7), with the actual solution of the Schrödinger equation ($n_x = 1$):

$$\varepsilon = \frac{h^2}{8ma^2}$$

Which yields the lower energy? Is this consistent with the variational principle—the principle that it is not possible for the calculated energy to be lower than the true energy for the system?

Exercises

1. Use the trial function (6) in Model 2 and the variational principle to obtain the value of $<\varepsilon>$ given in equation (7).

2. Calculate the energy, in joules, for the particle-on-a-line with the trial function in Model 2 and with the actual solution to the Schrödinger equation; let the length of the line = 1000 nm. Express the difference in the energy as a percent. Would you characterize Ψ_{trial} as: a good guess; a bad guess; an OK guess? Can you make a better guess?

Model 3: Trial Function for the Helium Atom.

- The Schrödinger equation cannot be solved for the helium atom.
- $He^{2+}(g) + 2\,e^- = He(g) \qquad \Delta U_{\text{experimental}} = -7{,}621$ kJ/mole (8)

We use a trial function that has each electron in a *1s* orbital.

$$\Psi_{\text{trial}} = \Psi_{\text{trial}}(1)\Psi_{\text{trial}}(2) = \Psi_{1s}(1)\,\Psi_{1s}(2) = \frac{1}{\pi}\left(\frac{Z}{a_o}\right)^3 e^{-Zr_1/a_o}\,e^{-Zr_2/a_o} \qquad (9)$$

where 1 and 2 refer to electron 1 and electron 2, respectively.

Critical Thinking Questions

8. The boundary condition for the electrons in an atom is that Ψ approaches zero as r approaches infinity. Does the trial function of Model 3 obey this boundary condition?

9. Provide a rationale for using the wave functions of the hydrogen atom as trial functions for the helium atom.

10. The trial probability function for electron 1 is $\Psi_{1s}^2(1)$. The trial probability function for electron 2 is $\Psi_{1s}^2(2)$. Suggest a reason why the full trial function is $\Psi_{1s}(1)\Psi_{1s}(2)$ and not $\Psi_{1s}(1) + \Psi_{1s}(2)$

Information

Note that it is possible to use the trial function (9) to calculate the average energy using different values for Z. Some results are shown in Table 1.

Table 1: Average Energy for Helium Atoms Using Trial Functions with Different Values of *Z*

Z	Energy (kJ/mole)
2.25	-6644
2.00	-7218
1.75	-7464
1.50	-7382
1.25	-6972

Critical Thinking Questions

11. Which value of Z in Table 1 provides a function which is the best approximation to the ground state of the He atom? What is the basis for your answer?

12. What physical explanation can you provide for the result that the "best" value of Z is NOT $Z = 2$?

Information

If Z, in Model 3, is treated as a variable, Z', the average energy for helium can be determined as a function of Z': $\varepsilon(\text{He}) = f(Z')$. The lowest energy can be minimized by setting the derivative of $<\varepsilon>$ with respect to Z' equal to zero and solving for the value of Z'.

$$\frac{d<\varepsilon>}{dZ'} = 0$$

The energy is minimized when $Z' = 27/16$; $<\varepsilon> = -7{,}475$ kJ/mole.

Exercises

3. When the parameter for the effective nuclear charge, Z', is used in equation (4), the following expression for the average value of the energy, $<\varepsilon>$, is obtained:

$$<\varepsilon> = (Z')^2 - 4Z' + \frac{5}{8}Z'$$

where Z' is the effective nuclear charge and the units are Hartree's, H. $H = 4.3597 \times 10^{-21}$ kJ.

(a) Show that application of the variational principle to this expression for $<\varepsilon>$ yields $Z' = 27/16$.

(b) Show that the energy calculated with this expression yields the value given above, $-7{,}475$ kJ/mole.

Model 4: Helium: Effective Nuclear Charge and Screening.

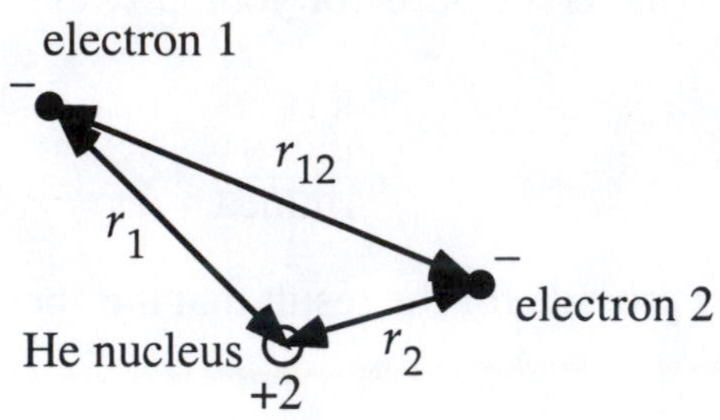

Critical Thinking Questions

13. a) What is the experimental value for the electronic energy of He?

    b) Are the average energies for the Ψ_{trial} with $Z' = 2$ and for the Ψ_{trial} with $Z' = 27/16$ higher or lower than the experimental electronic energy of helium?

14. Which is the better trial function for the helium atom: Ψ_{trial} with $Z' = 2$ or Ψ_{trial} with $Z' = 27/16$? Explain.

15. According to the figure in Model 4, electron 2 is closer to the nucleus than electron 1. In this configuration, electron 2 is said to *screen* the nuclear charge on electron 1. In models 1 and 2 both electrons are in a *1s* orbital. Do you think that electron 1 can screen the nuclear charge from electron 2?

16. Provide a rationale in terms of the physical system for why the trial with $Z' = 27/16$ gives a better energy value that when $Z' = 2$.

Information

- An alternative to the use of the term *screening* is the term *effective nuclear charge*. For He, the effective nuclear charge is 27/16.
- Even better values for the energy of He can be calculated by using trial functions that have more parameters. A trial function with 2000 parameters yields an energy value that is within the experimentally determined value for helium.

Model 5: Lithium.

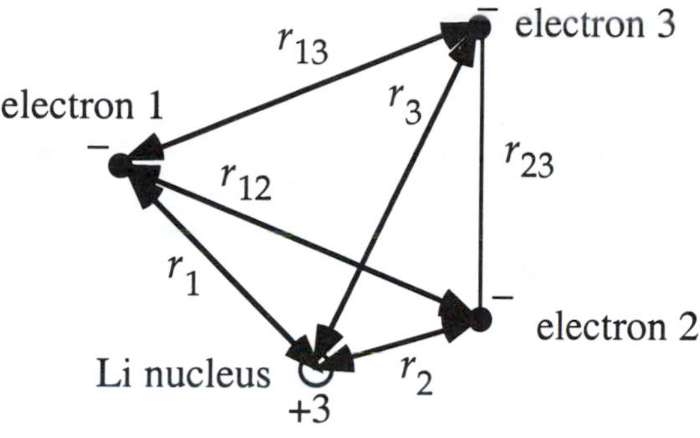

- The Schrödinger equation cannot be solved for the lithium atom.
- It has been determined from photoelectron spectroscopy (PES) that two of the three electrons are at the same energy level and the third electron is at a different (and higher) energy level. [Recall that PES measures the ionization energies of electrons in atoms (and molecules).]

$$\Psi_{\text{trial}} = \Psi_{\text{trial}}(1)\Psi_{\text{trial}}(2)\Psi_{\text{trial}}(3) = \Psi_{1s}(1)\Psi_{1s}(2)\Psi_{2s}(3) \qquad (10)$$

where 1, 2, and 3 refer to electron 1, electron 2, and electron 3, respectively.

Critical Thinking Questions

17. How many kinetic energy terms are present in the expression for the classical electronic energy of lithium? Hint, see equation (3) for helium.

18. How many <u>attractive</u> Coulombic energy terms are present in the expression for the classical electronic energy of lithium?

19. How many <u>repulsive</u> Coulombic energy terms are present in the expression for the classical electronic energy of lithium?

20. Write the Hamiltonian operator for lithium. Hint: see equation (3).

Model 6: Lithium: Effective Nuclear Charge and Screening.

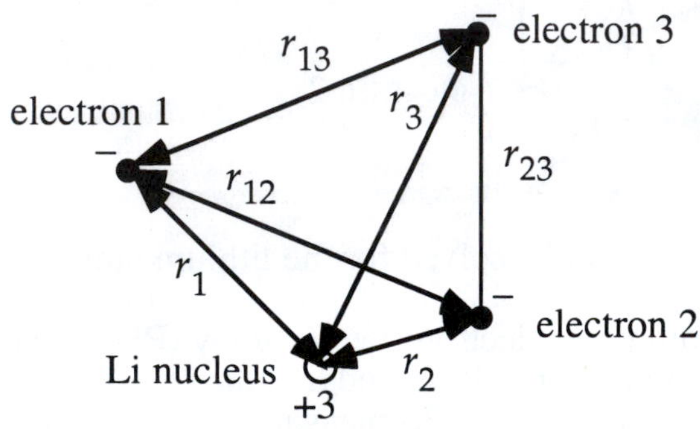

Critical Thinking Questions

21. When the electrons are arranged as in Model 6, electron 2 is closer to the nucleus than electron 3. In this configuration, electron 2 is said to *screen* the nuclear charge on electron 3. Sketch a similar diagram in which electron 2 screens electron 1.

22. Sketch a diagram in which electron 3 screens electron 1 and/or electron 2.

23. Based on the trial function (eq. 10) and your current understanding of the electronic structure of Li, explain why this configuration is less likely than the previous two.

24. Comment on the validity of this statement: To some extent, every electron shields the nuclear charge from every other electron.

25. The probability functions of Ψ_{2s} and Ψ_{2p} for an electron in a hydrogen atom are shown below. Recall that Coulomb's law is inversely proportional to distance. If electrons are present in Ψ_{1s}, as they are in lithium, provide a rationale for why an electron in a $2p$ orbital is more shielded from the nuclear charge than an electron in a $2s$ orbital.

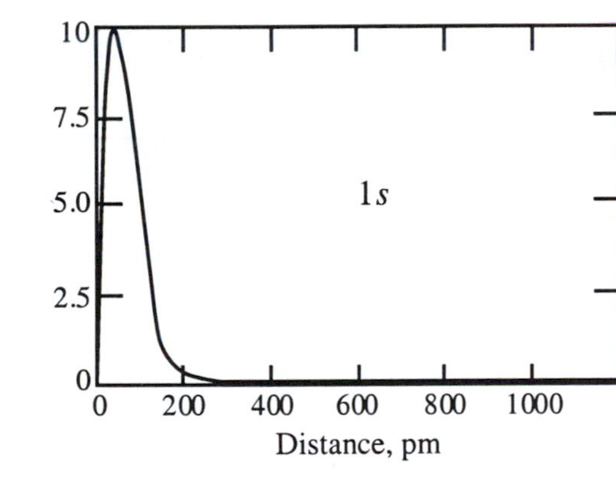

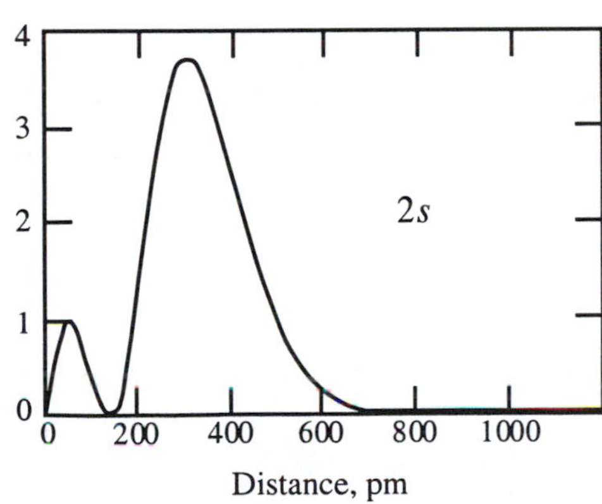

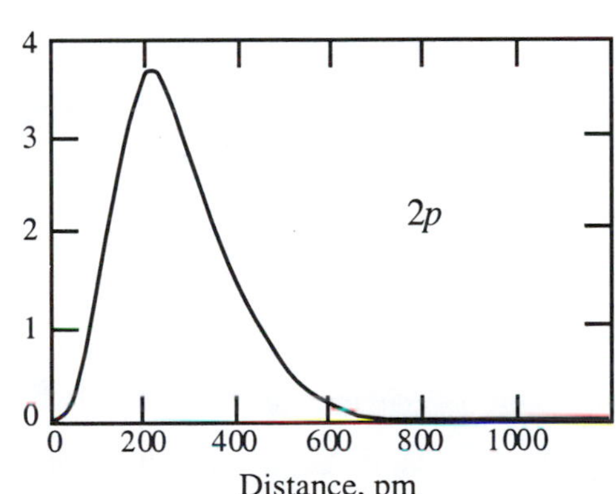

26. According to your answer above, which is the lower-energy electron configuration in lithium, $1s^2 2s^1$ or $1s^2 2p^1$? Explain why the ground state of lithium is $\Psi_{1s}(1)\Psi_{1s}(2)\Psi_{2s}(3)$ and not $\Psi_{1s}(1)\Psi_{1s}(2)\Psi_{2p}(3)$.

Exercises

4. Write the expression for the classical electronic energy of a Be atom. Write the Hamiltonian operator for a Be atom. Write a trial function for the electrons in a Be atom.

5. How many electron-electron repulsion terms are found in the Hamiltonian for the carbon atom? For the gold atom?

6. Write a trial function for the carbon atom. [Hint: see equation 10.]

ChemActivity 10*

Electron Spin

(When does the music stop?)

We have seen that three quantum numbers arise from the solution of the Schrödinger equation for the hydrogen atom: *n* (related to the energy of the electron); ℓ (related to the angular momentum of the electron; *m* (related to the z-component of the angular momentum of the electron. However, these three quantum numbers are not sufficient to explain all of the experimental properties of the hydrogen atom—specifically, spectral and magnetic properties. To correct this situation, George Uhlenbeck and Samuel Goudsmit, Dutch physicists, proposed that an electron spins about an axis (like a spinning top) and has a *z* component of angular momentum in addition to the angular momentum caused by its orbital motion. They called this additional angular momentum "**spin angular momentum**".

---

* Some material in this activity is taken from or based on that presented in CA 11 of R.S. Moog and J.J. Farrell, *Chemistry: A Guided Inquiry*, 2nd Edition, John Wiley & Sons, Inc. New York, 2002. This material is used by permission of John Wiley & Sons, Inc.

Model 1: Magnetic Moments

According to the classical laws of electricity and magnetism, a charged particle moving with angular momentum has a magnetic dipole moment. If the particle has a negative charge, an electron for example, the magnetic moment vector is in the opposite direction of the angular momentum vector.

$$\vec{L} = \sqrt{\ell(\ell+1)}\ \hbar$$

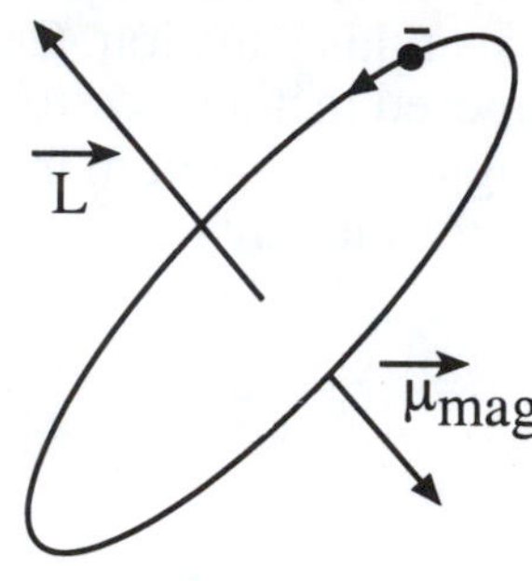

$$\vec{\mu}_{mag} \propto -\vec{L}$$

Critical Thinking Questions

1. Given the direction of the motion of the electron in Model 1, explain why the angular momentum vector is pointing in the direction shown.

2. Given the angular momentum vector in Model 1, explain why the magnetic moment vector is pointing in the direction shown.

3. According to the results of the hydrogen atom (Schrödinger equation), should an electron in a 1*s* orbital have a magnetic moment?

Information

One result of the solution of the Schrödinger equation for the one-electron atom is that the angular momentum of an electron in an *s* orbital is zero. Thus, an electron in an *s* orbital should not have a magnetic moment. Experimentally, however, we find that an electron in an *s* orbital does have a magnetic moment. To accommodate this experimental fact, a new feature must be added to the theory—electron spin. Recall that the angular momentum that arises from the solution of the Schrödinger equation is called **orbital angular momentum** and is characterized by the quantum numbers ℓ and m.

$$|\vec{L}| = \sqrt{\ell(\ell+1)}\,\hbar \tag{1}$$

$$L_z = m\,\hbar \tag{2}$$

The angular momentum that an electron has in addition to the orbital angular momentum is called the **spin angular momentum**. The equations for the spin angular momentum are analogous to the equations for orbital angular momentum.

$$|\vec{S}| = \sqrt{s(s+1)}\,\hbar \tag{3}$$

$$S_z = m_s\,\hbar \tag{4}$$

Experimentally, we find that the value of s in equation (3) is always 1/2, and that m_s can be either +1/2 or –1/2.

Therefore,

$$|\vec{S}| = \sqrt{\frac{3}{4}}\,\hbar \tag{5}$$

$$\text{and } S_z = \frac{1}{2}\hbar \quad \text{or} \quad S_z = -\frac{1}{2}\hbar \tag{6}$$

These electron spin values stem from a set of experiments based on a prototype experiment known as the Stern-Gerlach experiment. When a beam of atoms with one electron in an s orbital was passed through an inhomogeneous magnetic field (this is known as a Stern-Gerlach experiment), the beam splits into two components of equal intensity, but deflected in opposite directions. This implies that there are two equal and opposite magnetic moments possible for the electron in an s orbital, and that half of the atoms have one type and half of the atoms have the other type. The electrons giving rise to these moments are often referred to as "spin up" and "spin down."

When a beam of He atoms similarly undergoes a Stern-Gerlach experiment, the beam passes through without being deflected. This implies that there is no magnetic field associated with the He atoms, even though there are two electrons present. Thus, the two electrons in the atom must have opposite spins—one "up" and one "down"—which cancel each other out and provide no overall magnetic moment.

Model 2: The electron configurations of the ground states (lowest energy states) of several elements.

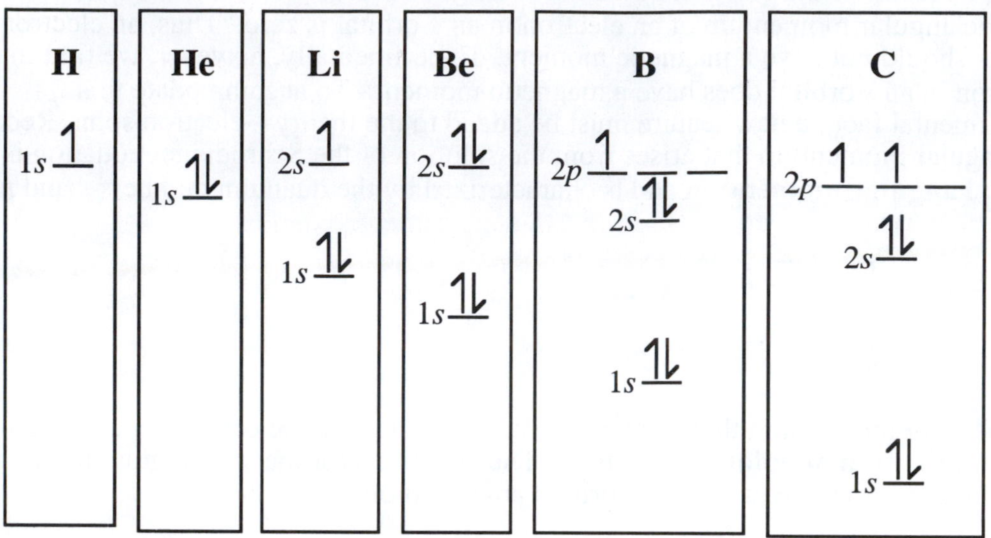

Relative energies shown are only approximate.

Critical Thinking Questions

4. What do the arrows in Model 2 represent?

5. What generalization can be made regarding the characteristics of the two electrons in a single "filled" *s* subshell?

6. For the ground state of carbon, the two electrons at the *2p*-energy-level were not placed in the same orbital; rather, they were placed in two different *2p* orbitals. Provide a physical basis to rationalize why this is expected to be the lower energy arrangement.

7. Predict the results of a Stern-Gerlach experiment on a beam of: a) Li atoms; b) Be atoms; c) B atoms; d) C atoms, e) N atoms. That is, predict whether the atoms will pass through undeflected or will be split into different components.

Information

An atom with an equal number of spin "up" and spin "down" electrons is known as **diamagnetic**, and the atom is repelled by a magnetic field. In this case we say that all of the electrons are "paired." If this is not the case—that is, if there are unpaired electrons—the atom is attracted to a magnetic field, and it is known as **paramagnetic**. The strength of the attraction is an experimentally measurable quantity known as the **magnetic moment.** The magnitude of the magnetic moment (measured in magnetons) is related to (but not proportional to) the number of unpaired electrons present. That is, the larger the number of unpaired electrons, the larger the magnetic moment.

Critical Thinking Questions

8. The magnetic moment of N(g) is larger than the magnetic moment of C(g). Why?

9. The magnetic moment of C(g) is about the same as the magnetic moment of O(g). Why?

10. Add the S_z components for the two electrons in He. What is $S_{z,\text{total}}$ for He?

11. Add the S_z components for the five electrons in B. What is $S_{z,\text{total}}$ for B?

12. Which of the following are diamagnetic? Paramagnetic?

    a) F(g)

    b) Ar(g)

    c) Na^+(g)

13. Based on Lewis structures, which of the following would you expect to be diamagnetic? Paramagnetic?

    a) $CH_4(g)$

    b) $NO_2(g)$

Information

The result that at most two electrons (of opposite spin) may occupy any atomic orbital has been formalized in a generalizaton known as the **Pauli Exclusion Principle**. It can be stated as follows:

No two electrons in an atom can have an identical set of quantum numbers (n, ℓ, m, m_s).

Another way of stating this is:

If two electrons in an atom have identical values of n, ℓ, and m, then they must have opposite spins.

Exercises

1. Using grammatically correct English sentences, describe the structure of a ^{13}C atom as completely as you can. Both the nucleus and the electrons should be considered in your description. You may use a diagram (or diagrams) as part of your answer, but you should explain the significance in words.

2. Which of the following are diamagnetic? Paramagnetic? S(g); Cl(g); $N_2(g)$; $F_2(g)$; benzene(g).

3. Add the S_z components for the 13 electrons in Al. What is $S_{z,total}$ for Al?

ChemActivity 11*

Photoelectron Spectroscopy

(What Is Photoelectron Spectroscopy?)

It turns out that the Schrödinger equation cannot be solved for any atom other than hydrogen. Thus, we rely on approximation methods for determination of the wavefunctions for all atoms other than hydrogen. Generally, we use the one-electron wavefunctions for all atoms and atomic ions as a starting point for these approximation methods. Terms such as *effective nuclear charge*, *nuclear screening*, and *electron-electron repulsion* are used to compensate for the fact that the energy levels of all subshells (within a shell) are not at the same energy level (as they are in the hydrogen atom). Furthermore, we recognize that each electron must be in an orbital and exist at a discrete, quantized energy level. Our knowledge of the actual values of these energy levels in atoms comes not from theory but from experimental evidence. An important method to obtain this evidence is **photoelectron spectroscopy** (PES).

---

* Some material in this activity is taken from or based on that presented in CAs 7-9 of R.S. Moog and J.J. Farrell, *Chemistry: A Guided Inquiry*, 2nd Edition, John Wiley & Sons, Inc. New York, 2002. This material is used by permission of John Wiley & Sons, Inc.

Information

For atoms with many electrons, we expect that the energy needed to remove an electron from the valence shell to be less than that needed to remove an electron from an inner shell because an inner shell is closer to the nucleus and is not as fully shielded as the outer valence electrons. Thus, for any given atom, less energy is needed to remove an electron from an $n = 2$ shell than from an $n = 1$ shell, and even less is needed to remove an electron from an $n = 3$ shell.

Figure 1. Each electron within an atom is found at a particular energy level.

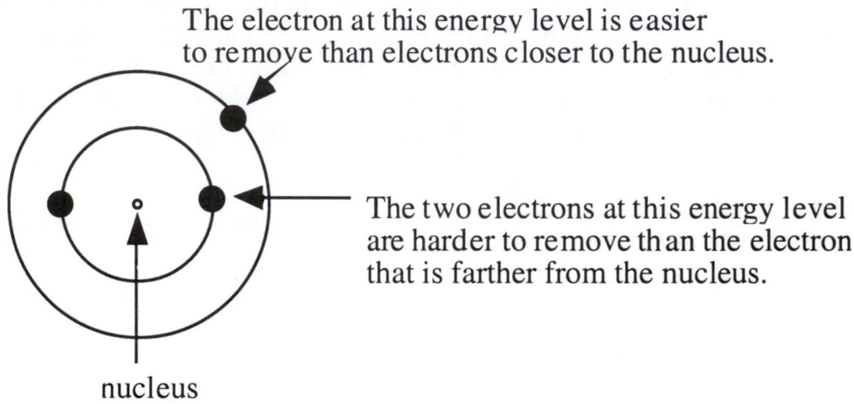

Photoelectron Spectroscopy (PES)

Ionization energies may be measured by the electron impact method, in which atoms in the gas phase are bombarded with fast-moving electrons. These experiments give a value for the ionization energy of the electron that is most easily removed from the atom—in other words, the first ionization energy for an electron in the highest occupied energy level. An alternative, and generally more accurate, method that provides information on all the occupied energy levels of an atom (that is, the ionization energies of all electrons in the atom) is known as photoelectron spectroscopy; this method uses photons to knock electrons out of atoms. Electrons obtained in this way are called photoelectrons.

Very high energy photons, such as very-short-wavelength ultraviolet radiation, or even x-rays, are used in this experiment. The gas phase atoms are irradiated with photons of a particular energy. If the energy of the photon is greater than the energy necessary to remove an electron from the atom, an electron is ejected with the excess energy appearing as kinetic energy, $\frac{1}{2}mv^2$, where v is the velocity of the ejected electron. In other words, the speed of the ejected electron depends on how much excess energy it has received. So, if IE is the ionization energy of the electron and KE is the kinetic energy with which it leaves the atom, we have

$$E_{photon} = IE + KE$$

or, upon rearranging the equation,

$$IE = E_{\text{photon}} - KE$$

Thus, we can find the ionization energy *IE* if we know the energy of the photon and we can measure the kinetic energy of the photoelectron. The kinetic energy of the electrons is measured in a photoelectron spectrometer.

If photons of sufficient energy are used, an electron may be ejected from *any* of the energy levels of an atom. Each atom will eject only one electron, but every electron in each atom has an (approximately) equal chance of being ejected. Thus, for a large group of identical atoms, the electrons ejected will come from all possible energy levels of the atom. Also, because the photons used all have the same energy, electrons ejected from a given energy level will all have the same energy. Only a few different energies of ejected electrons will be obtained, corresponding to the number of energy levels in the atom.

The results of a photoelectron spectroscopy experiment are conveniently presented in a *photoelectron spectrum*. This is essentially a plot of the number of ejected electrons (along the vertical axis) vs. the corresponding ionization energy for the ejected electrons (along the horizontal axis). It is actually the kinetic energy of the ejected electrons that is measured by the photoelectron spectrometer. However, as shown in the equation above, we can obtain the ionization energies of the electrons in the atom from the kinetic energies of the ejected electrons. Because these ionization energies are of most interest to us, a photoelectron spectrum uses the ionization energy as the horizontal axis.

Model 1: Photoelectron Spectroscopy.

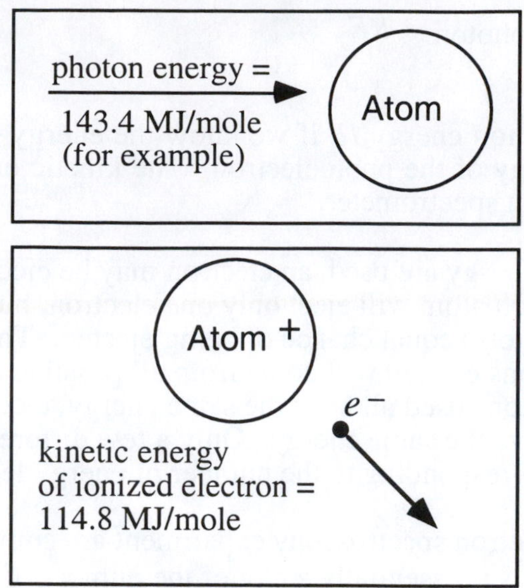

IE of electron = 28.6 MJ/mole

Critical Thinking Questions

1. Show that the *IE* of the electron in the model is 28.6 MJ/mole.

2. What is meant by the term "energy level"?

3. Based on Model 2 in CA 10, predict the number of peaks in a photoelectron spectrum of: a) H b) C c) Ne

Model 2. Simulated low resolution photoelectron spectra of helium and neon.

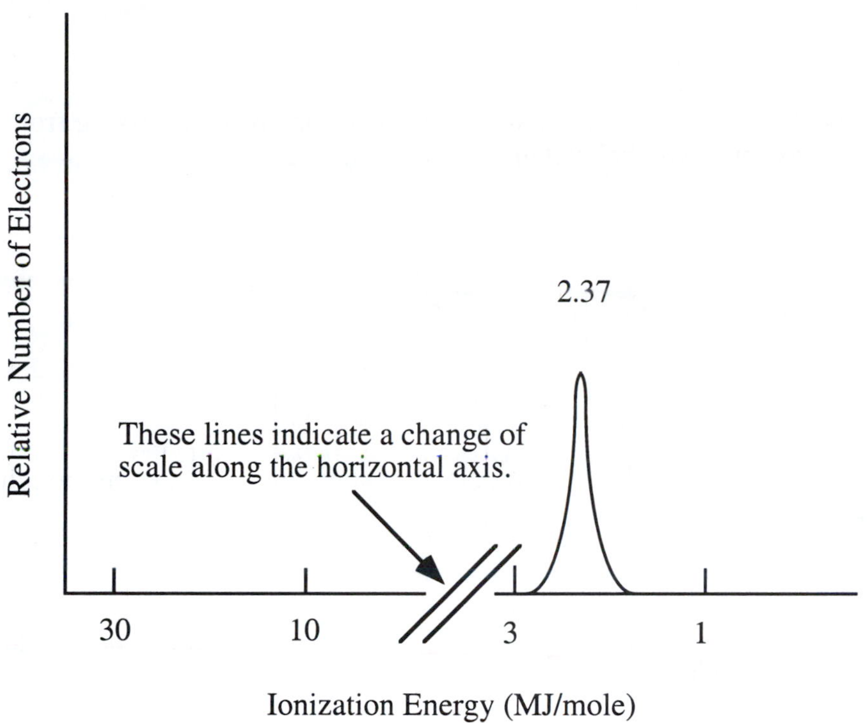

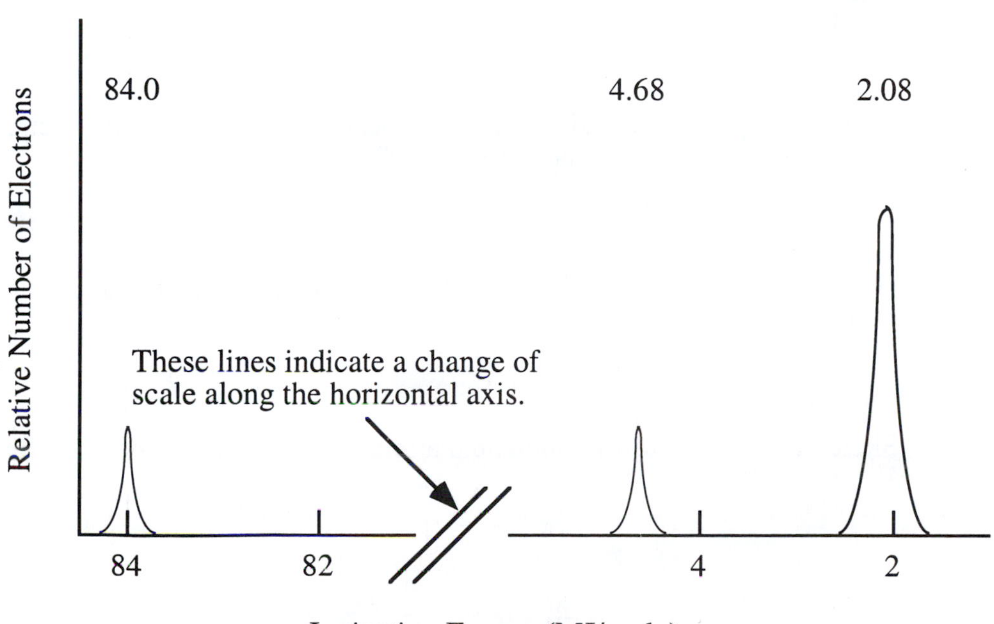

Critical Thinking Questions

4. Which spectrum in Model 2 is the spectrum of helium? Explain.

5. Index each peak in both spectra. That is, designate each peak as arising from *1s* or *2p*, or whatever orbital you believe gives rise to the peak.

6. What determines the height (or intensity) of each peak in a photoelectron spectrum?

7. Why is there only one peak in the photoelectron spectrum of He, even though there are two electrons in each atom?

8. A *hypothetical* atom in a galaxy far, far away has 2 electrons at one energy level and 3 electrons at another energy level (see energy level diagram below).

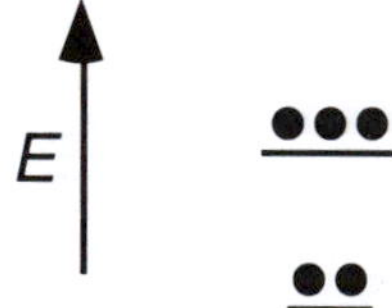

a) How many peaks (1,2,3,4, or 5) will appear in a photoelectron spectrum of a sample of this hypothetical atom? Why?

b) Describe the relative height of the peaks in the photoelectron spectrum of a sample of this hypothetical atom.

c) Sketch the PES of this hypothetical atom.

Model 3: Ionization Energies.

PES can be used to find the ionization energies of electrons in atoms and molecules. Some of the atomic ionization energies are shown in Table 1.

Table 1. Ionization energies (MJ/mole) for the first 18 elements.

Element	1*s*	2*s*	2*p*	3*s*	3*p*
H	1.31				
He	2.37				
Li	6.26	0.52			
Be	11.5	0.90			
B	19.3	1.36	0.80		
C	28.6	1.72	1.09		
N	39.6	2.45	1.40		
O	52.6	3.04	1.31		
F	67.2	3.88	1.68		
Ne	84.0	4.68	2.08		
Na	104	6.84	3.67	0.50	
Mg	126	9.07	5.31	0.74	
Al	151	12.1	7.19	1.09	0.58
Si	178	15.1	10.3	1.46	0.79
P	208	18.7	13.5	1.95	1.06
S	239	22.7	16.5	2.05	1.00
Cl	273	26.8	20.2	2.44	1.25
Ar	309	31.5	24.1	2.82	1.52

Critical Thinking Questions

9. Explain why the ionization energy of a 1*s* electron in F is greater than the ionization energy of a 1*s* electron in O.

10. For K, the first 18 electrons will fill the same orbitals as Ar. If the 19th electron of K is found in the 4*s* orbital, would the ionization energy be closest to 0.42, 1.4, or 2.0 MJ/mole? Explain. [Hint: compare to Na and Li.] Show a predicted photoelectron spectrum based on this assumption.

11. For K, if the 19th electron of K is found in the third subshell of $n = 3$, would the ionization energy be closest to 0.42, 1.4, or 2.0 MJ/mole? Explain. [Hint: Look for analogous patterns in the data.] Show a predicted photoelectron spectrum based on this assumption.

12. Based on your understanding of the electron configuration of K, which of the above predicted spectra (CTQ 10 or 11) is expected to be comparable to the experimental PES of K?

Exercises

1. Radiation from a helium discharge tube produces photon with a wavelength of 58.4 nm. If this radiation is used to ionize ground state hydrogen atoms, how much kinetic energy will be carried away by the electrons (assuming that the electron, not the proton, accounts for all of the kinetic energy carried away).

2. Use the values in Table 1 to sketch the PES of Al.

3. Indicate whether the following statement is true or false and explain your reasoning:

   The photoelectron spectrum of Mg^{2+} is expected to be identical to the photoelectron spectrum of Ne.

4. For which atom is the 1*s* energy level at a lower (more negative) energy, Cu or I? Why?

5. For which atom is the 2*p* energy level at a lower (more negative) energy, Pd or Gd? Why?

6. For which atom is the 1*s* energy level at a lower (more negative) energy: Cu; H, He; Pt; I? Why?

7. For which atom is the 1*s* energy level at a lower (more negative) energy, Cu or Cu^{2+}?

ChemActivity 12*

Electron Configurations

(Why is PES so useful?)

PES enables us to determine that electrons in a 2*p* orbital are not at the same energy as the electrons in a 2*s* orbital (for a multielectron atom). Experimental evidence (PES, electronic spectra, and magnetic properties of atoms) allows us to determine the electron configurations of atoms in spite of our theoretical shortcomings. This information is not only useful for a full description of the properties of atoms, but it is invaluable for a description of molecule formation and for the properties of molecules.

---

* Some material in this activity is taken from or based on that presented in CA 10 of R.S. Moog and J.J. Farrell, *Chemistry: A Guided Inquiry*, 2nd Edition, John Wiley & Sons, Inc. New York, 2002. This material is used by permission of John Wiley & Sons, Inc.

Model 1: Relative Energy Levels of Isolated H, He, and He^+ (1*s* through 4*p*).

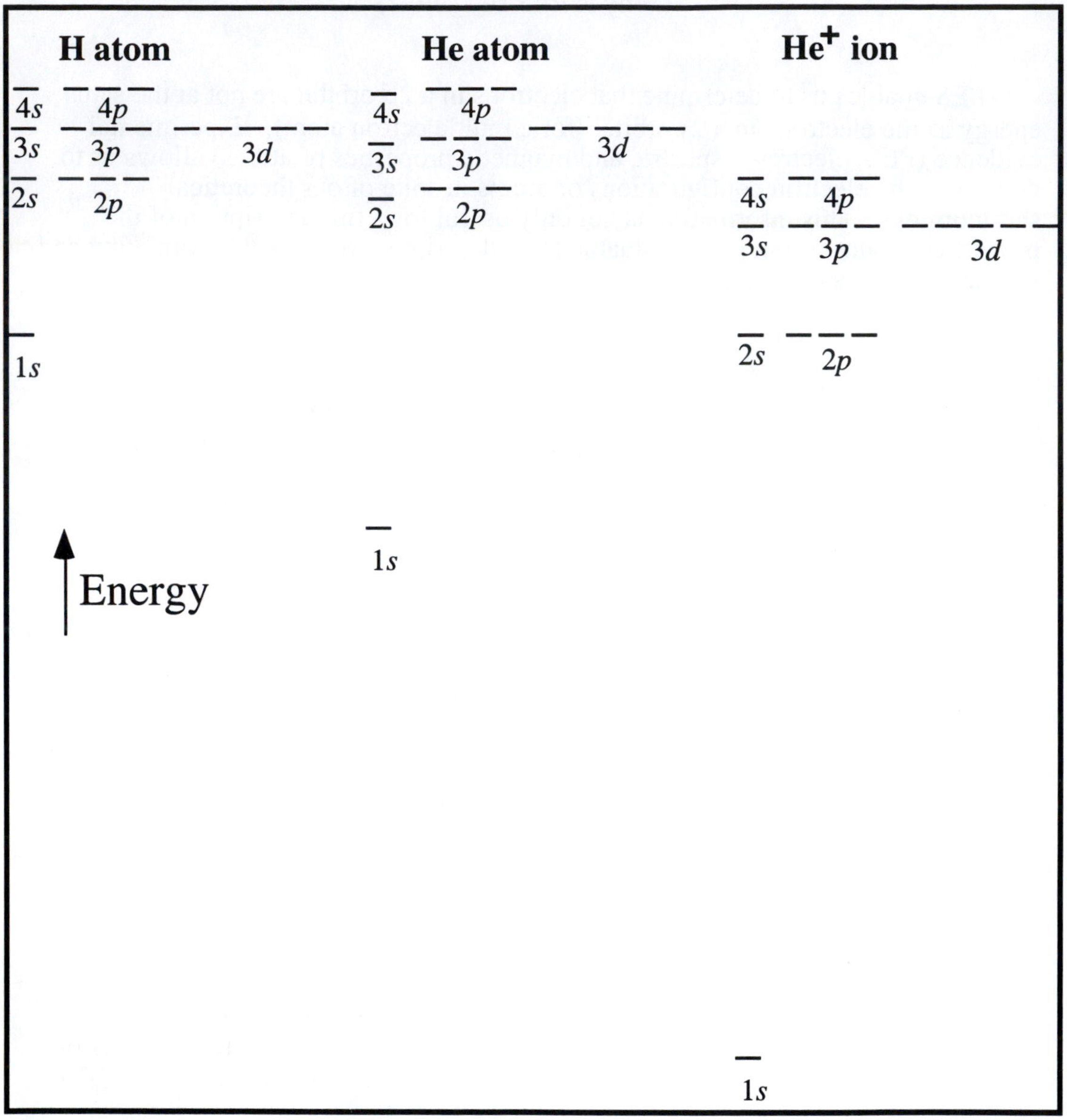

H atom

4*s* 4*p*

3*s* 3*p* 3*d*

2*s* 2*p*

1*s*

He atom

4*s* 4*p*

3*s* 3*p* 3*d*

2*s* 2*p*

1*s*

He^+ ion

4*s* 4*p*

3*s* 3*p* 3*d*

2*s* 2*p*

1*s*

Energy

Critical Thinking Questions

1. What is the nuclear charge of H? He? He^+?

2. How many electrons are there in H? He? He^+?

3. Why is the 1*s* energy level for a He atom lower than the 1*s* energy level for H?

4. Why is the 2*s* energy level of He at a lower energy than the 2*p* energy level of He? That is, why is the electron configuration $1s^12s^1$ lower in energy than the configuration $1s^12p^1$? [Hint: in which orbital is the electron more effectively screened from the nucleus? You may wish to refer to the figures in CTQ 25 of CA 9.]

5. The 3*s* energy level of He is shown to be at a lower energy than the 3*d* energy level of He. That is, the electron configuration $1s^13s^1$ is shown to be lower in energy than the configuration $1s^13d^1$. What inference can be drawn from this result concerning the relative shielding of electrons in the 3*s* and 3*d* orbitals by an electron in the 1*s* orbital?

6. Why are the 3*s* ,3*p*, and 3*d* energy levels of He^+ at the same energy?

Information

The 1*s* level in He is at a considerably lower energy than the 1*s* level in H because each electron in He only partially shields the other electron in He from the nuclear charge of +2; the effective nuclear charge on an electron in the 1*s* orbital of He is +1.69e. However, the 2*s* level in He is almost (but not quite, see Model 1) at the same energy as the 2*s* level in H.

Critical Thinking Question

7. Which is probably closer to the effective charge experienced by an electron in the 2*s* orbital of He in the configuration $1s^12s^1$, +2*e*; +1.9*e*; +1.5*e*; +1.1*e*? Explain your reasoning.

Exercises

1. Carefully explain why the 2*p* energy level in Li is at a higher energy than the 2*s* level in Li. That is, explain why the configuration $1s^22s^1$ is lower in energy than $1s^22p^1$.

2. Briefly explain why the 2*s* and 2*p* energy levels are at the same energy for Li^{2+}.

Information

For atoms (and ions) with large numbers of electrons, the ordering of the energy levels gets quite complicated. Increased nuclear charge results in a lowering of energy for all orbitals, but the magnitude of the effect varies depending on both *n* and ℓ. One reason for this is the variation in the effectiveness of the shielding for these different orbitals . For example, an electron in a 4*s* orbital is much less effectively screened from the nucleus than one in a 3*d* orbital because the 4*s* wavefunction has a much larger amplitude close to the nucleus. As atomic number increases, the energy of the 4*s* orbital is thus expected to become approximately equal to that of the 3*d* at some point.

Model 2: A Hypothetical Transition Metal

Consider a hypothetical atom for which the energies of the 4*s* and 3*d* levels are essentially identical.

—   — — — — —

4*s*   3*d*

Critical Thinking Question

8. Consider a situation in which two electrons must be placed in the diagram in Model 2. Rank the following arrangements of these electrons in order of increasing energy and explain your reasoning:
   a) $4s^2$
   b) $3d^2$ (same orbital)
   c) $3d^2$ (different orbitals)
   d) $4s^1\ 3d^1$

Information

When the 4*s* and 3*d* orbitals are close in energy, there are a variety of complicating factors related to interelectronic repulsion which make predicting the ground state electron configuration of these species difficult. For this reason, the only way to determine the ground state electron configuration for atoms (or ions) with large numbers of electrons (more than about 18) is through experimental measurements.

Model 3. Simulated low resolution photoelectron spectrum of potassium.

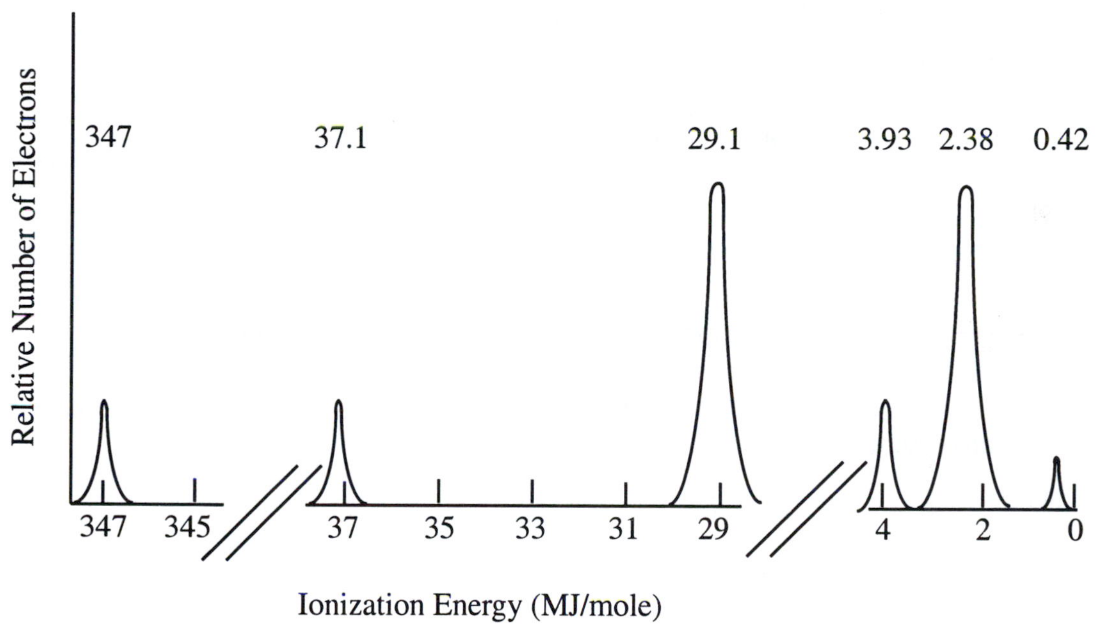

Critical Thinking Question

9. Compare the PES spectrum above to your predicted spectra in CTQs 10 and 11 from CA 11. According to the PES of potassium shown above, where is the 19th electron found in potassium? Explain.

Information

We can obtain photoelectron spectra for other atoms and atomic ions (cations and anions) and from these spectra the electron configurations of these species can be obtained. Other information about electron configuration of atoms and ions can be obtained from other experiments, including the measurement of magnetic moments and the observation of absorption and emission spectra in the absence and presence of magnetic fields. From these experiments, we have determined that the ground state configuration of V is [Ar]$4s^23d^3$, and for Mn it is [Ar]$4s^23d^5$, but the ground state configuration for Cr is [Ar]$4s^13d^5$. Experimental evidence also shows that the ground states of cations formed from transition metals generally have electrons in d orbitals, and not in the outermost s orbital of the parent atom.

Critical Thinking Questions

10. What is the expected electron configuration of V^{2+}?

11. The $4s$ level in Cr is at roughly the same energy as the $3d$ level. Explain how interelectronic repulsion can be used to rationalize the ground state electron configuration of Cr ($3d^5\ 4s^1$). What does this suggest about the lowest energy arrangement of electrons among degenerate orbitals?

Exercises

3. Give the electron configuration of each of the following: Al; Al^{3+}; Ni; Ni^{2+}; Zn; Zn^{2+}; Cl; Cl^-; Pd; Pd^{2+}; Gd; Gd^{3+}.

4. Add the S_z components for the 14 electrons in Si. What is $S_{z,\text{total}}$ for Si?

5. Add the S_z components for all of the electrons in Ni. What is $S_{z,\text{total}}$ for Ni?

6. Add the S_z components for all of the electrons in Ni^{2+}. What is $S_{z,\text{total}}$ for Ni^{2+}?

7. Add the S_z components for all of the electrons in Gd^{3+}. What is $S_{z,\text{total}}$ for Gd^{3+}?

ChemActivity 13

Term Symbols

(What is a term symbol?)

Atoms are characterized by the energy and the angular momentum of the electrons within the atom. The angular momenta (spin, orbital, and total) are often embodied in a symbol, called a **term symbol**. The term symbol is a useful concept and is ubiquitous in the literature of chemistry and physics.

Model 1: Experimentally obtained energy level diagram for a carbon atom with the electron configuration: $1s^2\ 2s^2\ 2p^2$.

		Energy Relative to Ground State (MJ/mole)
Level C	——— 0.260	
Level B	——— 0.122	
Level A	——— 0	

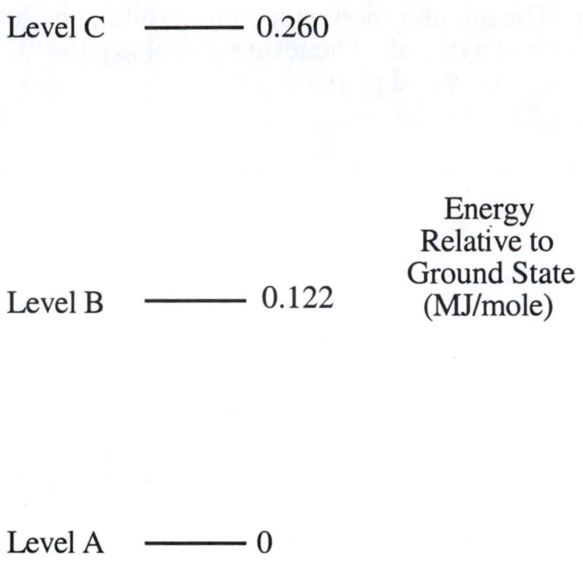

All three energy levels (A,B,C) have the electron configuration: $1s^2\ 2s^2\ 2p^2$.

Critical Thinking Questions

1. What arrangement of the two electrons in the three $2p$ orbitals do you expect results in the lowest energy (the ground state; energy level A)?

2. What factors can you think of which might cause other arrangements of the two electrons in the three $2p$ orbitals to have a higher energy?

3. What arrangement of electrons do you think is most likely to give rise to energy level C, the highest energy level?

Model 2: The 2p² electron configuration of Carbon.

There are 15 ways to place the two electrons into the three 2*p* orbitals. This corresponds to 15 **microstates** for the p^2 electron configuration. Three of the microstates are shown below. Not all 15 microstates are at the same energy level; on the other hand, they are not at 15 different energy levels either.

	↑	↑	—	$M_S = 1$	$M_L = 1$
	↓	—	↓	$M_S = -1$	$M_L = 0$
	↑↓	—	—	$M_S = 0$	$M_L = 2$
$m =$	1	0	−1		

M_S = the sum of the m_s for all of the electrons

M_L = the sum of the m for all of the electrons

$S_{z,\text{atom total}} = M_S \hbar$

$L_{z,\text{atom total}} = M_L \hbar$

Critical Thinking Questions

4. Show that the M_S and M_L values in Model 2 are correct.

5. Predict which state is expected to be the lower energy state, $M_S = 1$ and $M_L = 1$ or $M_S = 0$ and $M_L = 2$. Why?

6. What is the maximum value of M_S for two electrons in the p^2 configuration? What is the minimum value?

7. What is the maximum value of M_L for two electrons in the p^2 configuration? What is the minimum value?

Exercises

1. Prepare a table, similar to the one in Model 2, that shows all 15 microstates for the p^2 configuration. Give the M_S and M_L values for each microstate.

2. Prepare a table, similar to the one in Model 2, that shows the six microstates for the p^5 configuration. (fluorine, for example). Give the M_S and M_L values for each microstate.

Model 3: Total Angular Momentum and the z-component of the Total Angular Momentum.

Recall that the *z*-component of angular momentum is quantized. This is true for both the total orbital angular momentum, $\vec{L}$, and the total spin angular momentum, $\vec{S}$.

For a state with total orbital angular momentum quantum number *L*,

$$|\vec{L}| = \sqrt{L(L+1)}\ \hbar \quad \text{and } L_z = M_L \hbar$$

$$\text{where } M_L = L, L-1, L-2, \ldots, -L$$

For a state with total spin angular momentum quantum number S,

$$|\vec{S}| = \sqrt{S(S+1)}\ \hbar \quad \text{and } S_z = M_S \hbar$$

$$\text{where } M_S = S, S-1, S-2, \ldots, -S$$

Figure 1. Angular momentum vector diagrams for $L = 2$, $M_L = 2$ and $L = 2$, $M_L = 1$.

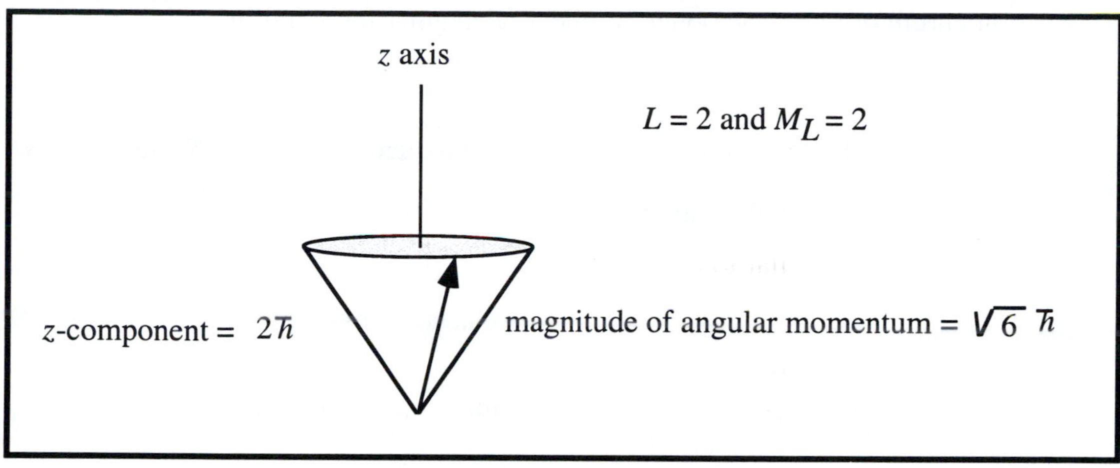

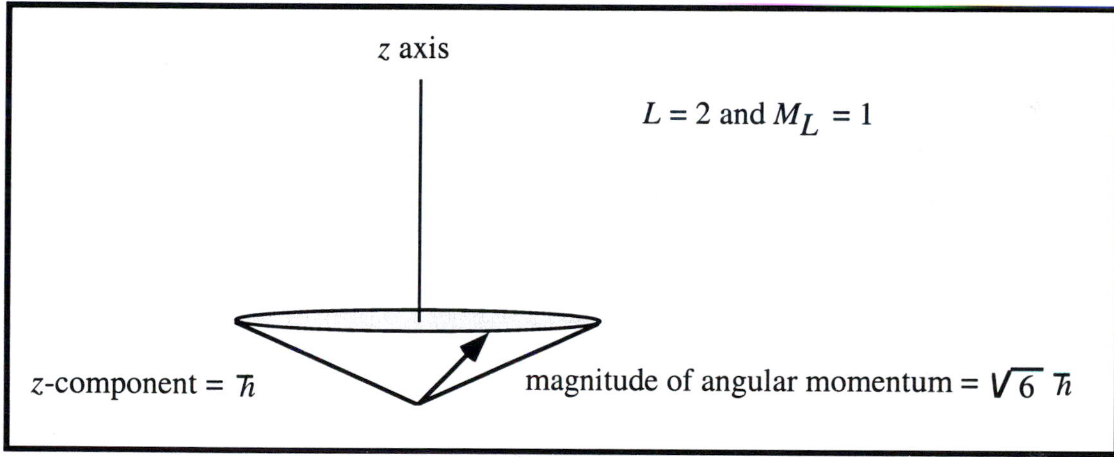

Information

It can be shown that $\Psi_{n\ell m}$ is not an eigenfunction of $\hat{L}_x$ or of $\hat{L}y$ (see Table 1 in CA 2). That is,

$$\hat{L}_x \Psi_{n\ell m} \neq \text{constant} \times \Psi_{n\ell m}$$

However, it can be shown that the average value of L_x and the average value of L_y is zero (see the Information section of CA 3). That is,

$$\int \Psi^*_{n\ell m} \hat{L}_x \Psi_{n\ell m} \, d\tau = 0$$

The above facts make sense if we conclude that the electron moves in a fashion such that the angular momentum vector precesses about the magnetic or electric field (the z-axis). In this case, the values of L_x and L_y keep changing, but the average values of L_x and L_y are zero and the z-component is a constant value.

Critical Thinking Questions

8. Examine Figure 1. Which is larger, the magnitude of the total orbital angular momentum or the *z*-component of the total angular momentum?

9. Suppose that the total spin angular momentum quantum number *S* equals 1. What is $|\vec{S}|$? What M_S values are possible?

10. Suppose that the total orbital angular quantum number *L* equals 1. What is $|\vec{L}|$? What M_L values are possible?

11. Suppose that the electrons in an atom have $S = 1$ and $L = 1$. What is $|\vec{S}|$? What is $|\vec{L}|$? Give the nine possible combinations of M_S and M_L. [These nine combinations represent nine microstates of the $S = 1$ and $L = 1$ quantum numbers.]

12. The maximum possible value of M_S for the $2p^2$ configuration is 1 and the minimum value is –1. Explain why this implies that the largest value of the quantum number *S* for the p^2 configuration is 1.

13. The maximum possible value of M_L for the $2p^2$ configuration is 2 and the minimum value is –2. Explain why this implies that the largest value of the quantum number *L* for the p^2 configuration is 2.

14. For the case in Figure 1 where $L = 2$ and $M_L = 2$, what is the value of θ (the angle between the angular momentum vector and the z-axis)?

Exercises

3. Suppose that the electrons in an atom have $S = 1/2$ and $L = 1$. What is $|\vec{S}|$? What is $|\vec{L}|$? What possible combinations of M_s and M_L are possible? How many states does this represent?

4. Suppose that the electrons in an atom have $S = 1$ and $L = 2$. What is $|\vec{S}|$? What is $|\vec{L}|$? What possible combinations of M_S and M_L are possible? How many states does this represent?

5. The angle between the angular momentum vector and the z-component of the angular momentum is θ.

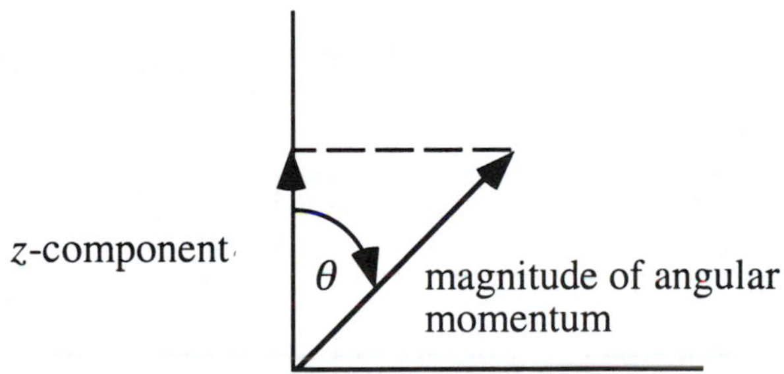

What is the value of θ for $L = 2$, $M_L = 2$ (see Figure 1)? What is the value of θ for $L = 2$, $M_L = 1$ (see Figure 1)?

6. What is the value of θ for $S = 4$, $M_S = 2$? Make a diagram, similar to Figure 1, for this combination of S and M_S.

Information

The electronic energy of an atom depends on a rather complicated array of factors: the kinetic energy of the electrons; the Coulombic attraction of the electrons for the nucleus; the Coulombic repulsion between the electrons; the magnetic interactions between electrons caused by the angular momentum of the electrons. All of these factors give rise to an energy for an atom that depends on its electronic configuration. For helium, the electron configuration $1s^1\ 2s^1$ is a higher energy state than $1s^2$. The lowest energy state is called the **ground state.** Higher energy states are called **excited states**. However, even within a given electronic configuration more than one energy level is often found. For example: for the p^2 configuration there are 15 microstates—not all of these microstates have the same energy. We use **term symbols** to differentiate the various energy levels that result from the same nominal electron configuration. The term symbols for a given electronic configuration depend on the orbital and spin quantum numbers as shown below in Model 4.

Model 4: Term Symbols.

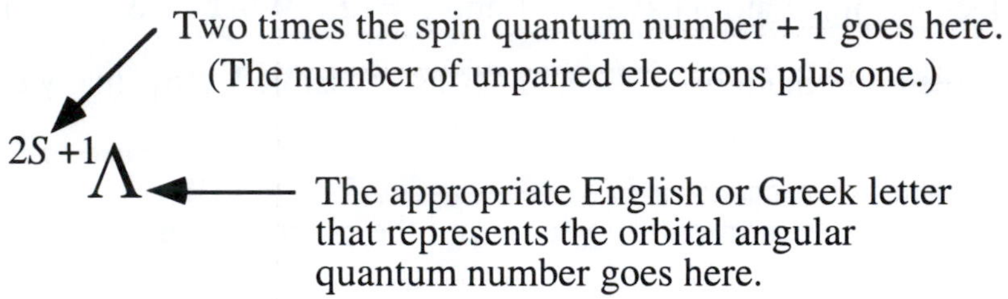

Table 1. Designations used to represent the orbital angular momentum in atoms and diatomic molecules.

Orbital Angular Momentum Quantum Number	One-Electron In An Atomic Orbital	Multielectron Atom	One-Electron In A Molecular Orbital	Multielectron Diatomic Molecule
0	*s*	*S*	σ	Σ
1	*p*	*P*	π	Π
2	*d*	*D*	δ	Δ
3	*f*	*F*		

Examples:

- The one-electron atomic orbital $\Psi(1,0,0)$ is called a $1s$ orbital because $n = 1$ and because the orbital angular quantum number is 0.
- The one-electron atomic orbital $\Psi(3,2,-1)$ is called a $3d$ orbital because $n = 3$ and because the orbital angular quantum number is 2.
- If the angular momentum of a multielectron atom is such that $S = 1$ and $L = 2$, the term symbol is ^3D and is spoken "triplet *D*."
- If the angular momentum of a multielectron atom is such that $S = 0$ and $L = 3$, the term symbol is ^1F and is spoken "singlet *F*."

Critical Thinking Question

15. What is the term symbol if the angular momentum of a multielectron atom is such that $S = 1$ and $L = 1$?

Exercise

7. Give the term symbol for each of the following combinations of S and L for a multielectron atom: $S = 0$ and $L = 0$; $S = 1$ and $L = 1$; $S = 1/2$ and $L = 2$; $S = 1$ and $L = 3$.

Model 5: Experimentally determined lower energy states of the carbon atom .

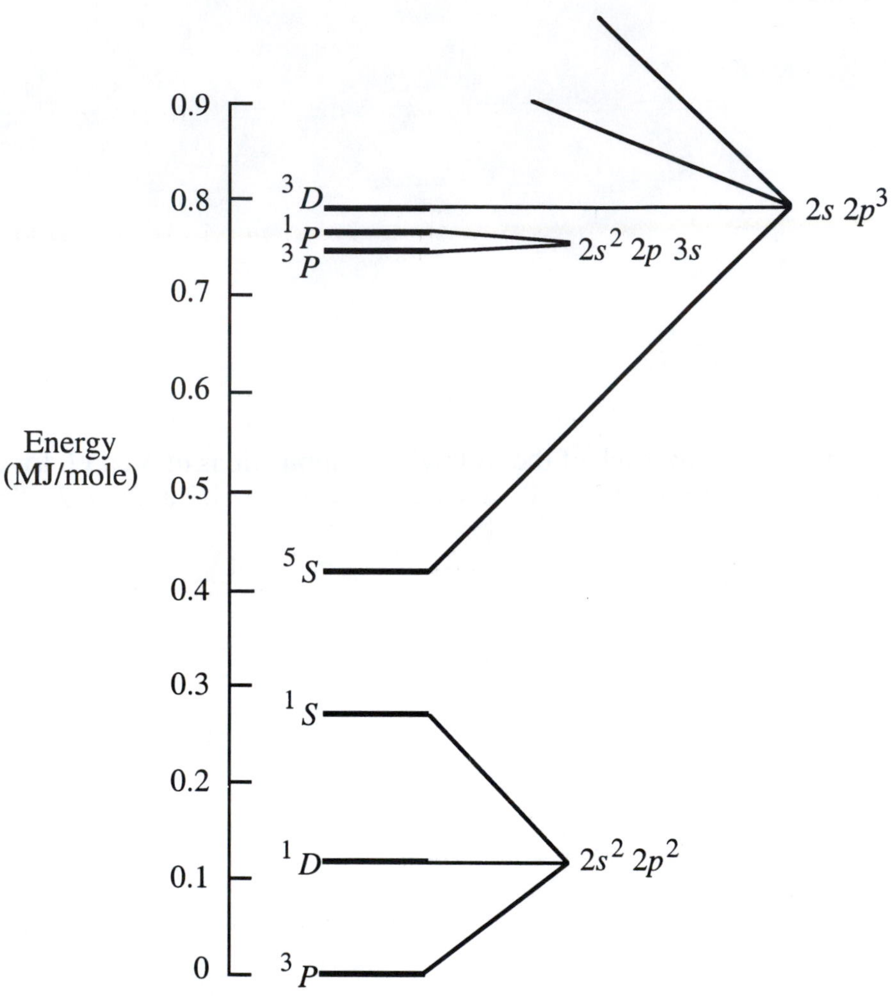

Critical Thinking Questions

16. According to Model 5, what is the term symbol for the ground state of carbon?

17. How many microstates are represented by the term symbol ^3P? [How many combinations of M_S, M_L are possible?]

18. How many microstates are represented by the term symbol ^1D?

19. How many microstates are represented by the term symbol ^1S?

20. Are the 15 microstates in the $2p^2$ configuration accounted for by the term symbols ^3P, ^1D, and ^1S?

Exercise

8. How many microstates exist for the electron configuration $2p^1\ 3s^1$ (an excited state of carbon)? According to Model 5, the ^3P and ^1P arise from this electron configuration. Do these terms symbols account for the appropriate number of states?

Model 6: Finding the Lowest-Energy Term Symbol.

Rule 1: For a given electron configuration, find the maximum value of M_S. This is the value of S for the ground state.

Rule 2: Retain the maximum value of M_S and find the maximum value of M_L. This is the value of L for the ground state.

Examples:

- The lowest energy electron configuration of nitrogen is $1s^2\ 2s^2\ 2p^3$.

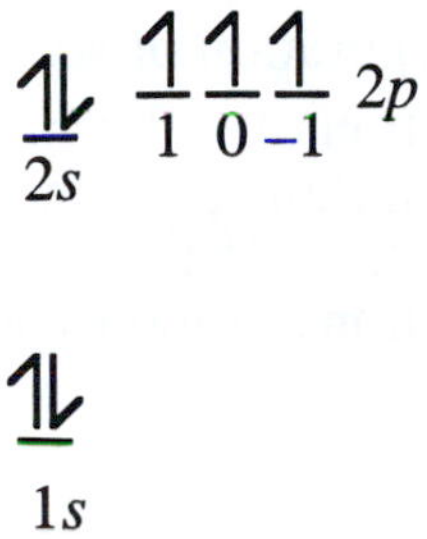

⇅ (2s) ↑ ↑ ↑ 2p (1 0 −1)

⇅ (1s)

The maximum value of S is 3/2. The maximum value of L is 0. (Note that L can be 2, two electrons in $m = 1$ and one electron in $m = 0$, but S would only be 1/2 in this case.) Thus, the lowest energy term symbol for this configuration is ^4S. This is the ground-state term symbol for nitrogen.

- An excited electron configuration of nitrogen is $1s^2\ 2s^1\ 2p^4$.

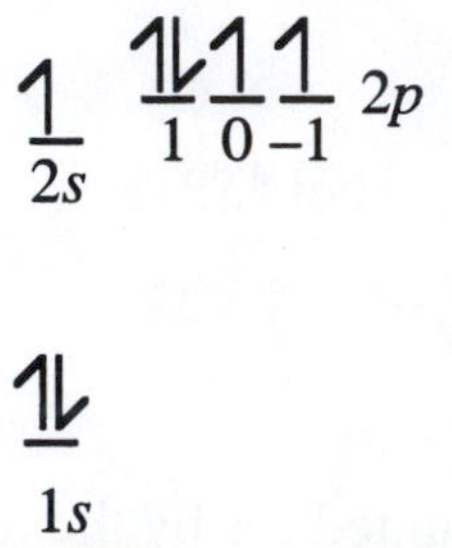

$$\underset{2s}{\uparrow} \qquad \underset{1}{\uparrow\downarrow}\ \underset{0}{\uparrow}\ \underset{-1}{\uparrow}\ 2p$$

$$\underset{1s}{\uparrow\downarrow}$$

The maximum value of S is 3/2. The maximum value of L is 1. (Note that L can be 2, two electrons in $m = 1$ and two electrons in $m = 0$, but S would only be 1/2 in this case.) Thus, the lowest energy term symbol for this configuration is ^4P. This is the lowest energy term symbol for <u>this</u> electron configuration of nitrogen. It is <u>not</u> the ground-state term symbol for nitrogen.

Critical Thinking Question

21. Why is the term symbol ^4P, above, not the ground-state term symbol for nitrogen?

Exercises

9. Find the ground-state term symbol for each of the following: He; O; N; Ne; Ti^{2+}; Ni; Ni^{2+}; Cu^{2+}; Cl^-; Ar ; Gd^{3+}.

10. Find the lowest energy term symbol for the $2s\ 2p^3$ configuration of carbon.

11. Find the lowest energy term symbol for the $3s^1$ configuration of sodium. Find the lowest energy term symbol for the $3s^0\ 3p^1$ configuration of sodium. Make an energy level diagram for these two term symbols. In a flame, ground sodium atoms get promoted to an excited state. The electron then returns to the ground state and emits a yellow photon. Use an arrow to indicate this transition on your energy level diagram.

ChemActivity 14

Hückel Molecular Orbitals

(Is HMO a health management organization?)

Although the Schrödinger equation can be written for molecules, the equation cannot be solved. Various approximation methods can be used to deal with our inability to solve the Schrödinger equation, and these methods differ greatly in the degree of sophistication and the need for known experimental values as parameters.

Most of these approximation methods have several connections to the hydrogen atom.

- The atomic orbitals (ao) on the atoms within the molecule are the one-electron atomic orbitals of the hydrogen. (For example, the atomic orbitals for the carbon atom are the $1s$, $2s$, and $2p$ one-electron orbitals.)
- These methods involve overlap of the atomic orbitals of the atoms within the molecule as the basis of the electron sharing (which reduces the nuclear/nuclear repulsion) and bond formation.
- These methods are, fundamentally, one-electron approximations. That is, these methods generate a one-electron molecular orbital (mo) energy diagram (note that the solution to the hydrogen atom generates a one-electron ao diagram). Then, the number of electrons in the molecule are placed into the one-electron mo energy diagram (similar to the placement of the electrons for multielectron atoms into the one-electron ao diagram for the hydrogen atom).

These are very severe approximations, and you might conclude that the results would be well off the mark. However, by including the values for known quantities (*IE*s, electronegativities, heats of formation, and so on) as parameters in semi-empirical equations some very useful information can be obtained.

One of the simplest approximation methods, and one that demonstrates many of the characteristics of more complicated methods, is known as the Hückel Molecular Orbital (HMO) method.

Model 1: The π-bond in Ethylene/HMO.

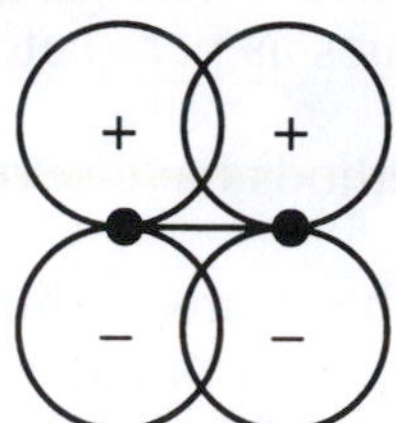

$2p$ orbital on carbon 1 = p_1 p_2 = $2p$ orbital on carbon 2

The classical expression for the energy and the Hamiltonian operator are never explicitly written in the HMO method. Furthermore, the trial wave function is written as a linear combination of atomic orbitals, LCAO, that comprise the π system. For ethylene, we use the $2p$ orbitals normal to the plane of the molecule.

$$\Psi_{\text{trial}} = c_1 p_1 + c_2 p_2 \qquad (1)$$

$p_1 \equiv \Psi_{2p}$ on atom 1 (the $2p$ orbital normal to the plane of the molecule)

$p_2 \equiv \Psi_{2p}$ on atom 2 (the $2p$ orbital normal to the plane of the molecule)

Only one-electron is described by Ψ_{trial} .

The coefficients c_1 and c_2 are constants, but their values are unknown at this point.

Now, the expression for the average value of the energy can be written:

$$<\varepsilon> = \frac{\int \Psi^*_{\text{trial}} \hat{H} \Psi_{\text{trial}} d\tau}{\int \Psi^*_{\text{trial}} \Psi_{\text{trial}} d\tau} \qquad (2)$$

Critical Thinking Questions

1. The two carbon nuclei repel each other. What location of an electron is most conducive to bond formation: when an electron is to the left of carbon 1; when an electron is between carbon 1 and carbon 2; when an electron is to the right of carbon 2? Explain.

2. In the HMO method, what is the trial function for the π-system of 1,3-butadiene?

Exercise

1. Write the HMO trial function for the π-system of benzene.

Information

The expression for the average value of the energy, equation (2) can be expanded:

$$<\varepsilon> = \frac{\int \Psi^*_{\text{trial}} \hat{H} \Psi_{\text{trial}} d\tau}{\int \Psi^*_{\text{trial}} \Psi_{\text{trial}} d\tau} = \frac{\int (c_1 p_1^* + c_2 p_2^*) \hat{H} (c_1 p_1 + c_2 p_2) d\tau}{\int (c_1 p_1^* + c_2 p_2^*)(c_1 p_1 + c_2 p_2) d\tau} \tag{3}$$

$$<\varepsilon> = \frac{\int c_1^2 p_1^* \hat{H} p_1 d\tau + \int c_2^2 p_2^* \hat{H} p_2 d\tau + \int c_1 c_2 p_1^* \hat{H} p_2 d\tau + \int c_2 c_1 p_2^* \hat{H} p_1 d\tau}{\int c_1^2 p_1^2 d\tau + \int c_2^2 p_2^2 d\tau + \int c_1 c_2 p_1^* p_2 d\tau + \int c_2 c_1 p_2^* p_1 d\tau} \tag{4}$$

$$<\varepsilon> = \frac{c_1{}^2 H_{11} + c_2{}^2 H_{22} + c_1 c_2 H_{12} + c_1 c_2 H_{21}}{c_1{}^2 + c_2{}^2 + c_1 c_2 S_{12} + c_1 c_2 S_{21}} \tag{5}$$

$$\text{where } H_{ij} = \int p_i^* \hat{H} p_j d\tau \tag{6}$$

$$S_{ij} = \int p_i^* p_j d\tau \tag{7}$$

The coefficients c_1 and c_2 can be treated as parameters, and the energy can be minimized with respect to each parameter. The resultant equations, (8) and (9) are known as secular equations.

$$\frac{\partial\langle\varepsilon\rangle}{\partial c_1} = 0 = (H_{11} - \langle\varepsilon\rangle S_{11})c_1 + (H_{12} - \langle\varepsilon\rangle S_{12})c_2 \quad (8)$$

$$\frac{\partial\langle\varepsilon\rangle}{\partial c_2} = 0 = (H_{21} - \langle\varepsilon\rangle S_{21})c_1 + (H_{22} - \langle\varepsilon\rangle S_{22})c_2 \quad (9)$$

To solve this system of equations, the integrals H_{ij} and S_{ij} must be evaluated.

Critical Thinking Questions

3. Why are only two secular equations obtained for the π system of ethylene?

4. If the trial function for a molecule contains four atomic orbitals, how many secular equations will be obtained?

5. Why is it desired to minimize the energy $\langle\varepsilon\rangle$ with respect to c_1 and c_2?

6. Why is $\frac{\partial\langle\varepsilon\rangle}{\partial c_i}$ set equal to zero?

7. For ethylene, do you expect $H_{11} = H_{22}$ and $S_{11} = S_{22}$? Explain your reasoning.

Model 2: Overlap of Two p Orbitals.

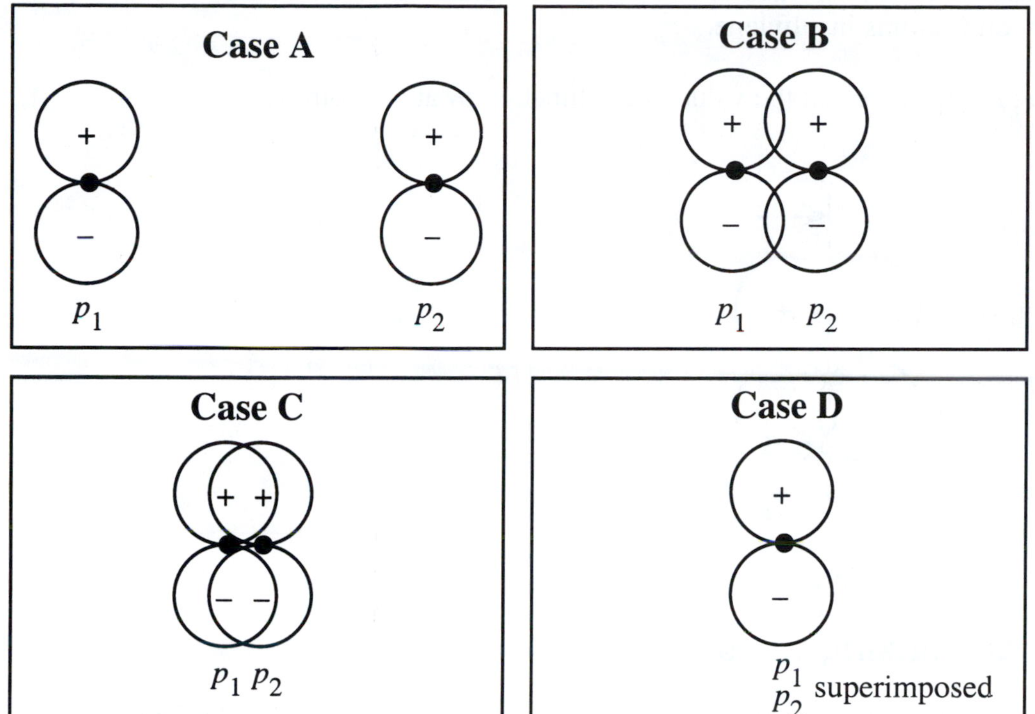

Critical Thinking Questions

8. Which case in Model 2 has:

   (a) the most overlap between the two atomic orbitals, p_1 and p_2?

   (b) the least overlap between the two atomic orbitals, p_1 and p_2?

9. Predict which case in Model 2 is most likely the best description for the π-system of the ethylene molecule.

Model 3: Overlap of p Orbitals.

Assume that the orbitals p_1 and p_2 are real functions, and that they represent p orbitals on C atoms in ethylene.

Let $p_1(y_d)$ represent the value of the function p_1 at the point y_d.

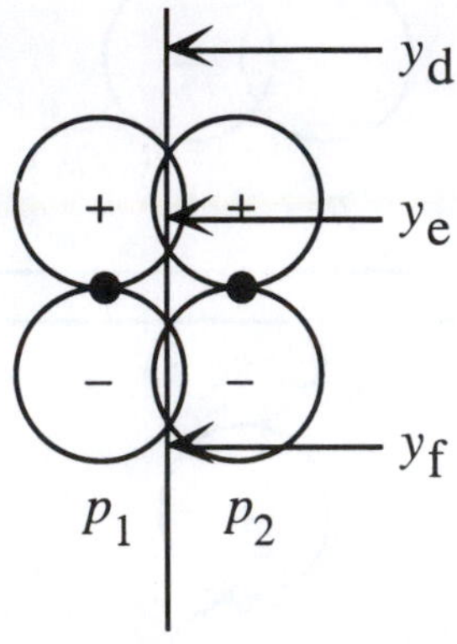

Critical Thinking Questions

10. For each case, indicate whether the value of the p orbital in Model 3 is positive, negative, or zero: $p_1(y_d)$; $p_1(y_e)$; $p_1(y_f)$; $p_2(y_d)$; $p_2(y_e)$; $p_2(y_f)$

11. Rank the following in terms of their relative magnitude and explain your reasoning: $p_1(y_d)$; $p_1(y_e)$; $p_1(y_f)$

12. Which has a greater magnitude): $p_1(y_e)$ or $p_2(y_e)$? Explain your reasoning.

13. For each case, indicate whether the value is positive, negative, or zero, and then rank them in order of increasing magnitude:
$p_1(y_d)\, p_2(y_d)$; $p_1(y_e)\, p_2(y_e)$; $p_1(y_f)\, p_2(y_f)$

14. Indicate the point y_g on the vertical line in Model 3 where the product $p_1(y_g)\, p_2(y_g)$ is a minimum. What is the value of the product at this point?

15. Explain why the integral S_{12} cannot be negative for ethylene.

16. Explain, in words, why $\int_{\text{all space}} p_1^* p_2 d\tau$ for Case A in Model 2 is one.

17. Explain, in words, why $\int_{\text{all space}} p_1^* p_2 d\tau$ for Case D in Model 2 is essentially zero.

18. Rank the values of S_{12} in increasing order for Cases A B, C, and D in Model 2.

Information

The integral $\int_{\text{all space}} p_1^* p_2 d\tau$ is called an *overlap integral*, S_{12}, because its magnitude is the fraction of overlap for the orbitals on carbon atoms 1 and 2. The overlap integral has a value between zero (the orbitals are infinitely apart) and one (complete overlap).

Model 4: The Assumptions of HMO.

- $S_{ij} = 0$ when $i \neq j$ (that is, all overlap integrals equal zero)
- $S_{ii} = 1$ (that is, all atomic orbitals are normalized)
- $H_{ij} = \beta =$ the exchange integral when i and j are adjacent atoms

  The exchange integral has no true classical analogy or interpretation. Examination of equation (6) reveals that this integral looks like an energy term that has the electron being shared by two different nuclei. When the two atoms are adjacent, we assume that the value of H_{ij} is sufficiently large that it cannot be ignored. We do not know the value of the exchange integral (at the moment). Later, we will determine the value of β from experimental evidence.

- $H_{ij} = 0$ when i and j are not adjacent atoms

  When the two atoms are not adjacent, we assume that the value of H_{ij} is sufficiently small that it can be set to zero.

- $H_{ii} = \alpha =$ the Coulomb integral

  Examination of equation (6) reveals that when $i = j$ this integral is an energy term corresponding to the electron being associated with one nucleus. We do not know the value of the Coulomb integral (at the moment). Generally, the Coulomb integral is viewed as the negative of the first *IE* of the isolated atom under consideration; for example, the first ionization energy of carbon is 1.09 MJ/mole and $\alpha = -1.09$ MJ/mole.

Both α and β have energy values that are negative.

Critical Thinking Questions

19. Use the assumptions of HMO to rewrite equations (8) and (9). Show that the result is:

$$0 = (\alpha - \varepsilon)c_1 + \beta c_2 \tag{10}$$

$$0 = \beta c_1 + (\alpha - \varepsilon)c_2 \tag{11}$$

where $\varepsilon = \langle\varepsilon\rangle$

These equations are also called the *secular equations*.

20. Solve these equations by the "brute force" technique. That is, solve for c_1 (in terms of α, β, ε, and c_2) with equation (10) . Use this value in equation (11) to solve for $<\varepsilon>$. [Hint: you will obtain two values for $<\varepsilon>$.]

21. Solve these equations by a "more elegant" technique. That is, solve for $<\varepsilon>$ by setting the determinant of the coefficients of the c_i values in equation (10) and (11) equal to zero. You should obtain the same values for $<\varepsilon>$.

$$\begin{vmatrix} (\alpha - \varepsilon) & \beta \\ \beta & (\alpha - \varepsilon) \end{vmatrix} = 0$$

Model 5: The Energy Levels of the π-bond of Ethylene.

There are two electrons in the π-bond of ethylene.

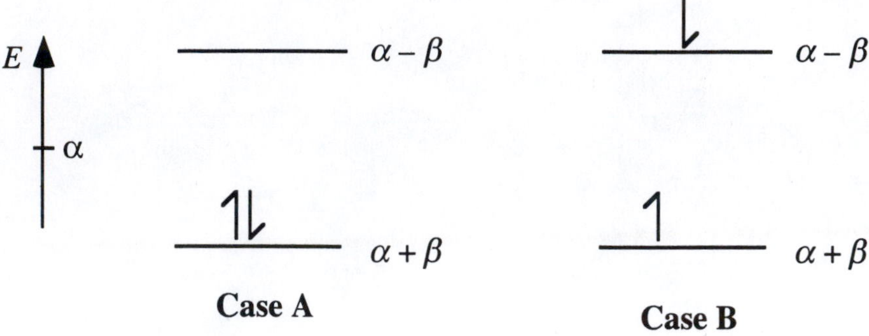

Critical Thinking Questions

22. Why is the energy level $\alpha + \beta$ at a lower energy than $\alpha - \beta$? [Hint: what is the sign of α? Of β?]

23. Which case (A or B) in Model 5 would be called the ground state? The excited state?

24. How much energy is required to move the electron from the ground state to the excited state (in terms of β)?

Exercise

2. The electronic transition from the ground to excited state in ethylene occurs at 165 nm. Calculate a value for β (in kJ/mole) based on the absorption in ethylene.

Information

The coefficients c_1 and c_2 are determined by using the normalization and orthogonality requirements of the wave functions. (Here we assume Ψ is real for simplicity.)

First, solve for the general relationship between c_1 and c_2.

$$\Psi = c_1 p_1 + c_2 p_2 \tag{12}$$

$$\int \Psi^2 d\tau = 1 = \int (c_1 p_1 + c_2 p_2)(c_1 p_1 + c_2 p_2) d\tau$$

$$1 = \int c_1^2 p_1^2 d\tau + \int c_2^2 p_2^2 d\tau + \int c_1 c_2 p_1 p_2 d\tau + \int c_1 c_2 p_2 p_1 d\tau$$

$$1 = c_1^2 + c_2^2 + c_1 c_2 S_{12} + c_1 c_2 S_{21} = c_1^2 + c_2^2$$

$$1 = c_1^2 + c_2^2 \quad (13$$

Next, solve for the coefficients of the wave function associated with $<\varepsilon> = \alpha + \beta$. Either one of the secular equations can be used—(10) or (11). Here, equation (10) is used.

$$0 = (\alpha - \varepsilon)c_1 + \beta c_2 = (\alpha - \alpha - \beta)c_1 + \beta c_2$$

$$\beta c_1 = \beta c_2$$

$$c_1 = c_2 \tag{14}$$

Combination of equations (13) and (14) yields:

$$c_1 = \frac{1}{\sqrt{2}} = c_2$$

Thus,

$$\varepsilon_\pi = \alpha + \beta, \ \Psi_\pi = \frac{1}{\sqrt{2}} p_1 + \frac{1}{\sqrt{2}} p_2 \tag{15}$$

Similarly, if $\varepsilon = \alpha - \beta$ (higher energy state), then $c_1 = -c_2$:

$$\varepsilon_{\pi^*} = \alpha - \beta, \ \Psi_{\pi^*} = \frac{1}{\sqrt{2}} p_1 - \frac{1}{\sqrt{2}} p_2 \tag{16}$$

Model 6: The Energy Levels and Orbitals of the π-bond of Ethylene.

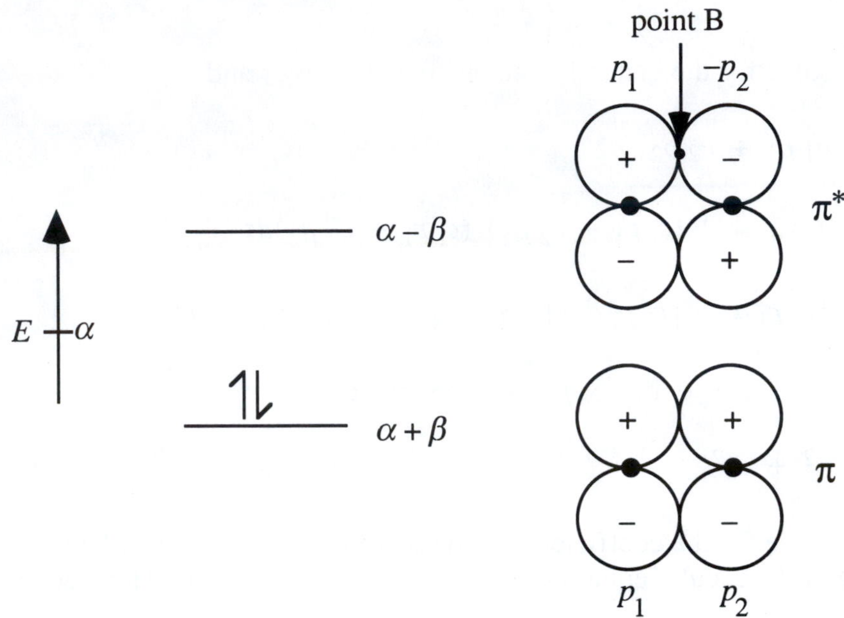

Critical Thinking Questions

25. For the π orbital, examine the wave function, equation (15), and the diagram in Model 6. Why are the two positive signs found on the same side of the molecular plane in Model 6?

26. For the π^* orbital, examine the wave function, equation (16), and the diagram in Model 6. Why are the two positive signs found on the opposite sides of the molecular plane in Model 6?

27. Point B is equidistant from the two carbon nuclei. Suppose that p_1 has a value of 4.2×10^{-6} pm$^{-3/2}$ at point B.

    a) What is the value of Ψ_{π^*} at point B?

    b) What is the value of Ψ_{π^*} at any point equidistant from the two nuclei?

28. How many nodes are there in Ψ_π ? In Ψ_{π^*} ? What is the relationship between the number of nodes and the energy of the mo?

Exercises

3. Show that Ψ_{π^*} is normalized.

4. Show that the wave functions for the π orbital and the π^* orbital are orthogonal.

ChemActivity 15

Conjugated π systems.

(Is delocalization a good thing?)

The basic concepts of HMO theory were developed in the consideration of ethylene: the **L**inear **C**ombination of **A**tomic **O**rbitals (LCAO); **secular equations**; the **Coulomb integral**; the **exchange integral**; the **overlap integral**. Now, we turn our attention to more complicated molecules, conjugated π systems.

Model 1: 1,3-butadiene.

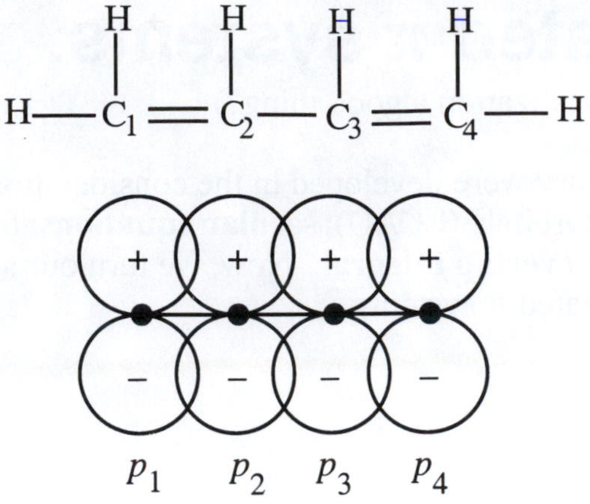

H—C$_1$=C$_2$—C$_3$=C$_4$—H

p_1 p_2 p_3 p_4

The trial function uses these four atomic orbitals.

$$\Psi_{\text{trial}} = c_1 p_1 + c_2 p_2 + c_3 p_3 + c_4 p_4 \tag{1}$$

As a result, there are four coefficients in the trial function. The trial function must be minimized with respect to each of the four coefficients.

$$\frac{\partial\langle\varepsilon\rangle}{\partial c_1} = 0 = (H_{11} - \langle\varepsilon\rangle S_{11})c_1 + (H_{12} - \langle\varepsilon\rangle S_{12})c_2 + (H_{13} - \langle\varepsilon\rangle S_{13})c_3 + (H_{14} - \langle\varepsilon\rangle S_{14})c_4$$

$$\frac{\partial\langle\varepsilon\rangle}{\partial c_2} = 0 = (H_{21} - \langle\varepsilon\rangle S_{21})c_1 + (H_{22} - \langle\varepsilon\rangle S_{22})c_2 + (H_{23} - \langle\varepsilon\rangle S_{23})c_3 + (H_{24} - \langle\varepsilon\rangle S_{24})c_4$$

$$\frac{\partial\langle\varepsilon\rangle}{\partial c_3} = 0 = (H_{31} - \langle\varepsilon\rangle S_{31})c_1 + (H_{32} - \langle\varepsilon\rangle S_{32})c_2 + (H_{33} - \langle\varepsilon\rangle S_{33})c_3 + (H_{34} - \langle\varepsilon\rangle S_{34})c_4$$

$$\frac{\partial\langle\varepsilon\rangle}{\partial c_4} = 0 = (H_{41} - \langle\varepsilon\rangle S_{41})c_1 + (H_{42} - \langle\varepsilon\rangle S_{42})c_2 + (H_{43} - \langle\varepsilon\rangle S_{43})c_3 + (H_{44} - \langle\varepsilon\rangle S_{44})c_4$$

Four secular equations are generated.

$$
\begin{aligned}
0 &= (\alpha - \varepsilon) c_1 + \beta c_2 + 0 + 0 \\
0 &= \beta c_1 + (\alpha - \varepsilon) c_2 + \beta c_3 + 0 \\
0 &= 0 + \beta c_2 + (\alpha - \varepsilon) c_3 + \beta c_4 \\
0 &= 0 + 0 + \beta c_3 + (\alpha - \varepsilon) c_4
\end{aligned}
$$

There are four solutions for the energy.

$$
\begin{aligned}
\varepsilon_1 &= \alpha + 1.6180\,\beta \\
\varepsilon_2 &= \alpha + 0.6180\,\beta \\
\varepsilon_3 &= \alpha - 0.6180\,\beta \\
\varepsilon_4 &= \alpha - 1.6180\,\beta
\end{aligned}
$$

Four wave functions are generated.

$$
\begin{aligned}
\Psi_1 &= 0.3717\,p_1 + 0.6015\,p_2 + 0.6015\,p_3 + 0.3717\,p_4 \\
\Psi_2 &= 0.6015\,p_1 + 0.3717\,p_2 - 0.3717\,p_3 - 0.6015\,p_4 \\
\Psi_3 &= 0.6015\,p_1 - 0.3717\,p_2 - 0.3717\,p_3 + 0.6015\,p_4 \\
\Psi_4 &= 0.3717\,p_1 - 0.6015\,p_2 + 0.6015\,p_3 - 0.3717\,p_4
\end{aligned}
$$

There is said to be a **conservation of orbitals**. The number of molecular orbitals generated is equal to the number of atomic orbitals used in the trial function.

A sketch of Ψ_1 is shown below.

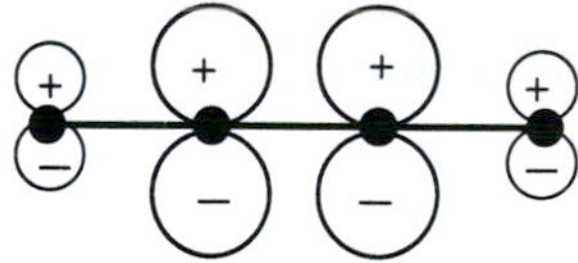

Critical Thinking Questions

1. Which energy level, $\mathcal{E}_i$, is the lowest energy level?

2. Which orbital, Ψ_i , is represented below. Explain.

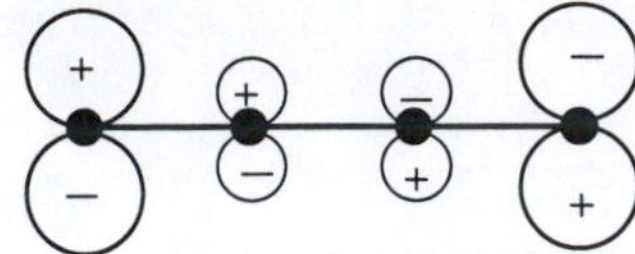

3. In CTQ 2, why is the *p* orbital on carbon 1 depicted larger than the *p* orbital on carbon 2?

4. If an electron is placed into Ψ_2 , is the electron more likely to be found on carbon atom 1 or carbon atom 2? Explain.

5. The magnitude of the coefficients on carbon atoms 1 and 4 have the same value in any one molecular orbital, Ψ_i . Why must this be the case?

6. Ψ_1 does not have any nodes (other than the node in the molecular plane). How many nodes does Ψ_2 have? Ψ_3 ? Ψ_4 ?

7. What is the apparent relationship between the number of nodes of the orbital and the energy level of the orbital?

8. For Ψ_1, there appears to be bonding over all four carbon atoms. For Ψ_2, there appears to be bonding carbon atoms 1 and 2, bonding between carbon atoms 3 and 4, and a node between carbon atoms 2 and 3. What type of bonding or nodes seem to be evident between the carbon atoms when an electron is described by Ψ_4?

Exercises

1. How many atomic orbitals should be used in the trial function for the π system of benzene? How many molecular orbitals would be generated?

2. How many atomic orbitals should be used in the trial function for the π system of naphthalene? How many molecular orbitals would be generated?

Model 2: MO Diagram of the π System of 1,3-butadiene.

Ψ_4 ——— $-1.6180\,\beta$

E ↑

Ψ_3 ——— $-0.6180\,\beta$

α

Ψ_2 ——— $0.6180\,\beta$

Ψ_1 ——— $1.6180\,\beta$

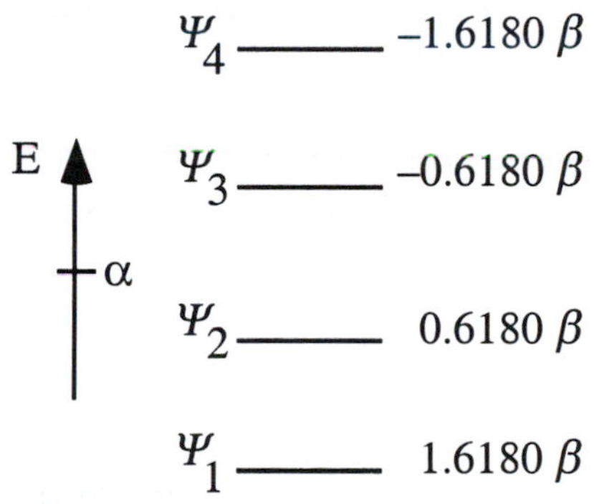

Critical Thinking Questions

9. What is the energy separation between Ψ_1 and Ψ_2? Ψ_2 and Ψ_3? Ψ_3 and Ψ_4?

10. How many electrons are found in the π system of 1,3-butadiene?

11. The electron configuration of the ground state of 1,3-butadiene can be written as $\Psi_1^2\Psi_2^2$. Write the electron configuration of the lowest-energy excited state of 1,3-butadiene?

Model 3: π-electron Charge Densities and π-electron Charges.

The square of a coefficient of an atomic orbital is taken as the fraction of an electron found near that atom (by each electron in an occupied molecular orbital). That is, for one electron in the molecular orbital

$$\Psi_2 = \quad 0.6015\, p_1 + \quad 0.3717\, p_2 - \quad 0.3717\, p_3 - \quad 0.6015\, p_4$$

0.3618 of the electron is found at atom 1, 0.1382 at atom 2, and so on. Thus, the π-electron charge density (a misnomer), q, at atom r is given by

$$q_r = \sum_i n_i \,(c_{ir})^2 \tag{2}$$

where n_i is the number of electrons in the *i*th molecular orbital , c_{ir} is the coefficient of atom *r* in the *i*th molecular orbital, and the sum is over all molecular orbitals containing one or more electrons.

The π-electron charge (also a misnomer) on atom *r* is taken as the number of electrons that an atom contributes to the π system minus the charge density, q_r. For example, if a carbon atom contributes one electron to the π system and has a π-electron charge density of 1.023, then the charge on the carbon atom is –0.023. On the other hand, if a pyrrole-like nitrogen contributes two electrons to the π system and has a π-electron charge density of 1.988, then the charge on the nitrogen atom is +0.012.

Critical Thinking Questions

12. Give a rationale for taking the square of a coefficient of an atomic orbital as the fraction of an electron found near that atom (by each electron in an occupied molecular orbital)?

13. What is the sum of the squares of all of the coefficients for Ψ_2? Explain why this makes sense.

14. Why does the number of electrons in the orbital, n_i, appear in equation (2)?

Exercises

3. Refer to Model 2 and CTQ 11. Calculate the π-electron charge density on atom 1, q_1, for the ground state of 1,3-butadiene. Calculate the π-electron charge density on atom 2, q_2. Without doing any calculations, what is the π-electron charge density on atom 3? On atom 4?

4. According to your energy level diagram for 1,3-butadiene, what is the $\Delta\varepsilon$ for the lowest energy electronic transition (in terms of β)? Use your value of β, calculated in CA 14, Exercise 2 to determine the wavelength of this transition. The experimental value for 1,3-butadiene is 217 nm. How well does your calculated value compare to the experimental value?

5. The lowest energy resonance structure and numbering system for imidazole is shown below:

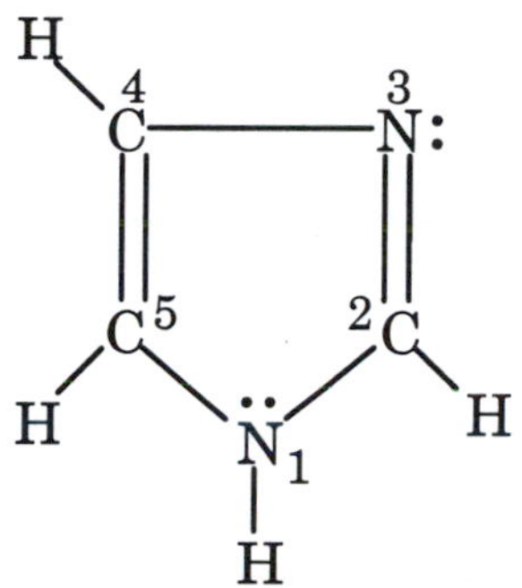

   a. Draw two resonance structures for imidazole that show that the lone-pair electrons on N1 (the pyrrole-type nitrogen) <u>are</u> contributors to the π system. (These same resonance structures will show that the lone-pair electrons on N3 (the pyridine-type nitrogen) are not contributors to the π system.)

   b. Give the total number of electrons in the π-system of imidazole. How many electrons are contributed by N1? By C2? By N3? By C4? By C5?

   c. Give the total number of p-orbitals in the π-system of imidazole. How many orbitals are contributed by N1? By C2? By N3? By C4? By C5?

   d. The HMO energy levels for the π-system of imidazole are:

   $\varepsilon_1 = \alpha + 2.3737\,\beta$

   $\varepsilon_2 = \alpha + 1.3785\,\beta$

   $\varepsilon_3 = \alpha + 0.6736\,\beta$

   $\varepsilon_4 = \alpha - 0.9586\,\beta$

   $\varepsilon_5 = \alpha - 1.4672\,\beta$

   Make a π-system energy level diagram for imidazole and place the correct number of electrons into the diagram.

e. The HMO wave functions for the π-system of imidazole are:

$$\Psi_1 = 0.6911\, p_1 + 0.3907\, p_2 + 0.3746\, p_3 + 0.3112\, p_4 + 0.3640\, p_5$$
$$\Psi_2 = 0.6055\, p_1 - 0.1168\, p_2 - 0.6454\, p_3 - 0.4502\, p_4 + 0.0248\, p_5$$
$$\Psi_3 = 0.0826\, p_1 + 0.5354\, p_2 + 0.2945\, p_3 - 0.4842\, p_4 - 0.6207\, p_5$$
$$\Psi_4 = 0.3835\, p_1 - 0.6700\, p_2 + 0.3354\, p_3 + 0.1808\, p_4 - 0.5087\, p_5$$
$$\Psi_5 = 0.0428\, p_1 + 0.3133\, p_2 - 0.4939\, p_3 + 0.6583\, p_4 - 0.4720\, p_5$$

A sketch of the top-view of Ψ_1 is shown below:

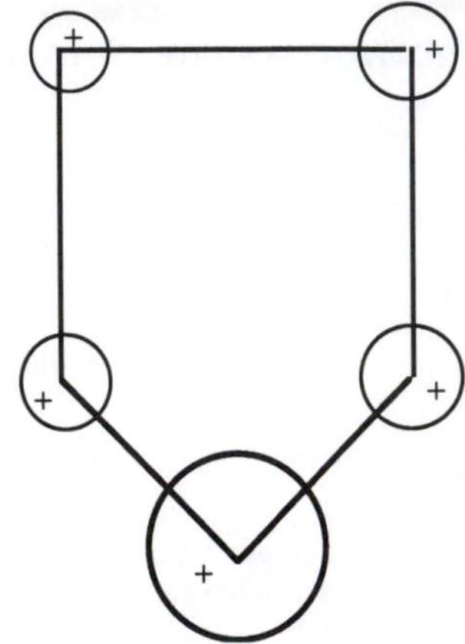

Make a sketch of the top view of the other four molecular orbitals.

f. If an electron is placed into Ψ_1, which atom "has" most of the electron? Which atom "has" least of the electron?

g. Calculate the π-electron charge density on atom 2 for the ground state configuration.

h. Calculate the π-electron charge on atom 2 for the ground state configuration.

Model 4: π-electron Bond Orders.

Bonding takes place when there is a significant sharing of electron density between two atoms. For an occupied orbital, the electron sharing is large when the wave function is large between the adjacent atoms. Thus, the bond order between adjacent atoms is considered to be large when the coefficients on the adjacent atoms are large and have the same sign. The π-electron bond order is another defined quantity. For adjacent atoms r and s, the bond order is given by

$$p_{rs} = \sum_{i} n_i \, (c_{ir} \, c_{is}) \tag{3}$$

where c_{ir} and c_{is} are the coefficients of atom *r* and *s*, respectively, in the *i*th molecular orbital, and the sum is taken over all orbitals.

Critical Thinking Questions

15. Suppose the coefficients for the *p* orbitals on two adjacent atoms are 0.50 and 0.01, respectively. Suppose the coefficients for the p orbitals between a second set of adjacent atoms is 0.25 and 0.26. Give a rationale for taking the products of the coefficients rather than the sum of the coefficients as a measure of the bond strength.

16. Suppose the coefficients for the *p* orbitals on two adjacent atoms are –0.25 and –0.26. Give a rationale for taking the products of the coefficients rather than the sum of the coefficients as a measure of the bond strength.

17. Suppose the coefficients for the *p* orbitals on two adjacent atoms are 0.25 and –0.26. Does this represent bonding between the two atoms, or will there be a node between the two atoms?

18. Why does the number of electrons in the orbital, n_i, appear in equation 3?

Exercises

6. For the ground state of 1,3-butadiene, calculate the π-electron bond order p_{12}. Calculate the π-electron bond order p_{23}. Without doing any calculations, give the π-electron bond order p_{34}.

7. What is the sum of the number of π bonds in the lowest energy resonance structure of 1,3-butadiene? What is the sum of the number of π bonds in the HMO description of 1,3-butadiene? That is, what is the sum of p_{12}, p_{23}, p_{34}? Explain why the amount of π-bonding is greater using HMO than with the lowest energy resonance structure.

8. Calculate the π-electron bond order between atom 2 and atom 3 in imidazole.

ChemActivity 16

Molecular Orbitals and Energies from MOPAC/AM1 .

(Is MOPAC related to Huckel?)

More sophisticated approximation methods for the determination of energy levels in molecules use all of the valence orbitals in the molecule (not simply the overlapping *p* orbitals used by HMO). The atomic orbitals on the atoms within the molecule are the one-electron atomic orbitals of the hydrogen. (That is, the atomic orbitals for carbon atom are the 1*s*, 2*s*, and 2*p* one-electron orbitals.)

These methods depend heavily on various known parameters: actual values of atomic energy levels for each atom; electronegativities; polarizabilities; formal charge; and so on. On the other hand, the results of the calculations are more accurate and more extensive than the HMO results. For example, these methods yield bond orders, charges on atoms, heats of formations, actual energies (in eV or kJ/mole), dipole moments, and other parameters.

MOPAC (Molecular Orbital Package) is a widely used semi-empirical molecular orbital package for the study of chemical structures and reactions. Semi-empirical Hamiltonians such as MNDO, MINDO/3, AM1, and PM3 are used to generate molecular orbitals for a given geometry. The molecular orbitals are used to calculate a heat of formation and the derivative of the heat of formation with respect to the molecular geometry. The geometry is varied to minimize the heat of formation (to find the most negative value). The MOPAC results can be used to calculate vibrational spectra, thermodynamic quantities, isotopic substitution effects, partition functions, heat capacities, and force constants. MOPAC can also be used to study chemical reactions.

The MOPAC calculations presented here are from calculations using CAChe software, from Oxford Molecular.

Model 1: The Energy Levels of Ethylene using MOPAC/AM1.

The molecular plane is the *xy*-plane

Coordinate system on each atom

$$\Psi_{\text{trial}} = c_1\, 2s_{C1} + c_2\, 2px_{C1} + c_3\, 2py_{C1} + c_4\, 2pz_{C1} + c_5\, 2s_{C2} + c_6\, 2px_{C2} + c_7\, 2py_{C2} + c_8\, 2pz_{C2} + c_9\, 1s_{H3} + c_{10}\, 1s_{H4} + c_{11}\, 1s_{H5} + c_{12}\, 1s_{H6} \quad (1)$$

Table 1. MOPAC/AM1 Energy Levels for Ethylene.

Energy Level	Energy (eV)	Energy Level	Energy (eV)
1	-33.16	7	1.44
2	-21.89	8	4.01
3	-15.80	9	4.40
4	-14.30	10	5.08
5	-11.84	11	5.56
6	-10.55	12	5.79

Figure 1. MOPAC/AM1 Energy Levels for MOs of Ethylene and Energy Levels of the Valence Orbitals of Carbon and Hydrogen.

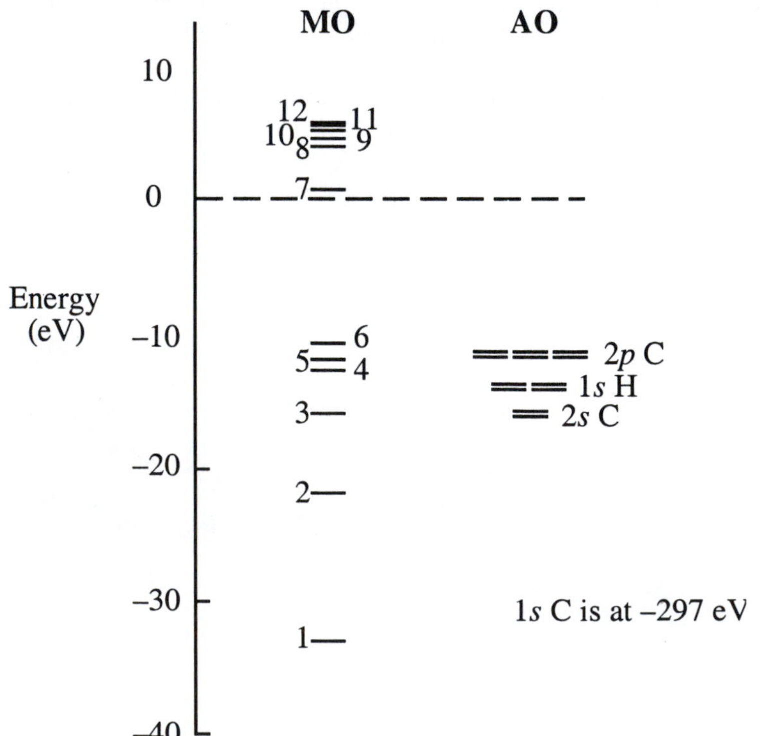

Table 2. MOPAC/AM1 Wave functions for Ethylene.

Atomic Orbital	Coefficients for each MO (Ψ_i) Ψ_1	Ψ_2	Ψ_3	Ψ_4	Ψ_5	Ψ_6
2*s*C1	-0.6229	0.4840	0.0000	0.0028	0.0000	0.0000
2*px*C1	-0.1784	-0.2911	0.0000	-0.6010	0.0000	0.0000
2*py*C1	0.0000	0.0000	-0.5427	0.0000	0.4630	0.0000
2*pz*C1	0.0000	0.0000	0.0000	0.0000	0.0000	0.7071
2*s*C2	-0.6229	-0.4840	0.0000	0.0028	0.0000	0.0000
2*px*C2	0.1784	-0.2911	0.0000	0.6010	0.0000	0.0000
2*py*C2	0.0000	0.0000	-0.5427	0.0000	-0.4629	0.0000
2*pz*C2	0.0000	0.0000	0.0000	0.0000	0.0000	0.7071
1*s*H3	-0.2002	-0.3009	-0.3205	0.2634	-0.3779	0.0000
1*s*H4	-0.2002	-0.3009	0.3205	0.2634	0.3779	0.0000
1*s*H5	-0.2002	0.3009	0.3205	0.2633	-0.3779	0.0000
1*s*H6	-0.2002	0.3009	-0.3205	0.2633	0.3779	0.0000

Atomic Orbital	Coefficients for each MO (Ψ_i) Ψ_7	Ψ_8	Ψ_9	Ψ_{10}	Ψ_{11}	Ψ_{12}
2*s*C1	0.0000	0.0000	0.4077	-0.3348	0.0000	-0.3155
2*px*C1	0.0000	0.0000	-0.1410	0.3270	0.0000	-0.6288
2*py*C1	0.0000	0.4533	0.0000	0.0000	-0.5345	0.0000
2*pz*C1	0.7071	0.0000	0.0000	0.0000	0.0000	0.0000
2*s*C2	0.0000	0.0000	0.4077	-0.3348	0.0000	0.3155
2*px*C2	0.0000	0.0000	-0.1410	-0.3270	0.0000	-0.6288
2*py*C2	0.0000	0.4533	0.0000	0.0000	0.5345	0.0000
2*pz*C2	-0.7071	0.0000	0.0000	0.0000	0.0000	0.0000
1*s*H3	0.0000	-0.3838	0.3962	0.3748	-0.3273	0.0505
1*s*H4	0.0000	0.3838	0.3962	0.3748	0.3274	0.0505
1*s*H5	0.0000	0.3838	0.3962	0.3748	-0.3274	-0.0505
1*s*H6	0.0000	-0.3838	-0.3962	0.3748	0.3273	-0.0505

Table 3. MOPAC/AM1 Atom Charges for Ethylene.

Atom	Atom Charge
C1	-0.2180
C2	-0.2180
H3	0.1090
H4	0.1090
H5	0.1090
H6	0.1090

Table 4. MOPAC/AM1 Bond Length and Bond Angles for Ethylene.

Atom Number (I)	Chemical Symbol	Bond Length (A) NA:I	Bond Angle (°) NB:NA:I	NA	NB
1	C				
2	C	1.326		1	
3	H	1.098	122.72	2	1
4	H	1.098	122.70	2	1
5	H	1.098	122.71	1	2
6	H	1.098	122.72	1	2

Figure 2. MOPAC/AM1 Bond Length and Bond Angles for Ethylene.

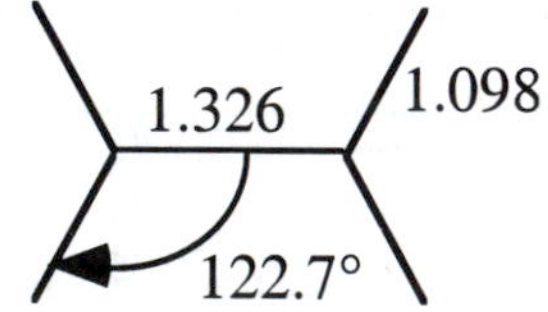

Table 5. MOPAC/AM1 Bond Orders for Ethylene.

C1–C2	2.002
C1–H5	0.958

Table 6. Other MOPAC/AM1 Values for Ethylene.

Heat of Formation	=	16.47 kcal/mole
Dipole Moment	=	0.00 D
Total Energy	=	-310.36 eV
Electronic Energy	=	-737.52 eV
Core-Core Repulsion	=	427.12 eV
Ionization Potential	=	10.55 eV
No. of Filled Levels	=	6
Molecular Weight	=	28.054

Critical Thinking Questions

1. How many atomic orbitals are used in the trial function (MOPAC) for ethylene?

2. Why aren't the two 1*s* orbitals on the carbon atoms used in the trial function?

3. What is the value (in eV) of the lowest energy molecular orbital of ethylene?

4. What is the value (in eV) of the highest energy molecular orbital of ethylene?

5. How many valence electrons are found in the molecule?

6. In the ground state, what is the highest occupied molecular orbital—the *HOMO*?

7. Why is the ionization potential of the molecule listed as 10.55 eV in Table 6?

8. In the ground state, what is the lowest unoccupied molecular orbital—the *LUMO*?

9. What is the value (in eV) of the lowest energy electronic transition for ethylene?

10. What is the wavelength, in nm, of a photon that promotes an electron from the *HOMO* to the *LUMO*? (1 eV = 1.6022×10^{-19} J) How well does this value compare to the experimental value of 165 nm?

11. Examine the molecular orbitals and find the MO that most resembles a π bond between carbon atoms 1 and 2.

12. Examine the molecular orbitals and find the MO that most resembles a π^* bond between carbon atoms 1 and 2.

13. According to Table 3, the carbon atoms are negatively charged and the hydrogen atoms are positively charged. Does this agree with tabulated values for the electronegativities of these elements?

14. What is the MOPAC/AM1 value for the sum of the partial charges on the atoms in ethylene. [Hint: see Table 3.] Does this value make sense?

15. What is the MOPAC/AM1 value for the carbon-carbon bond length in ethylene? [Hint: see Table 4.]

16. What is the MOPAC/AM1 value for the C–H bond length in ethylene?

17. What is the MOPAC/AM1 value for ∠HCC in ethylene?

18. What is the MOPAC/AM1 value for ∠HCH in ethylene?

19. What is the MOPAC/AM1 value for the C1–H5 bond order? [Hint. See Table 5.] What is the C2–H3 bond order?

Exercises

1. A sketch of MO1 for ethylene is shown below. In MO1, which two atomic orbitals are responsible for the bulk of the bonding?

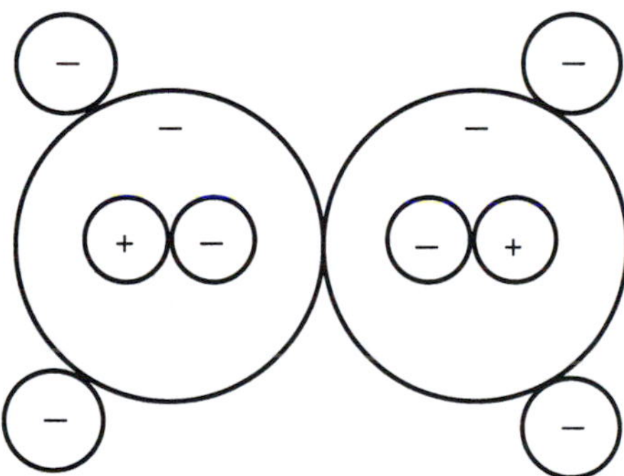

Examine MO4. Which two atomic orbitals are responsible for the bulk of the bonding in MO4?

2. How many valence orbitals are in benzene? How many molecular orbitals and energy levels should a MOPAC calculation of benzene yield?

3. How many valence orbitals are in ethane? How many molecular orbitals and energy levels should a MOPAC calculation of ethane yield?

4. How many valence orbitals are in CCl_4? How many molecular orbitals and energy levels should a MOPAC calculation of CCl_4 yield?

5. The MOPAC secular matrix for ethylene is a 12 x 12 matrix. What is the dimension of the secular matrix for benzene? For ethane? For naphthalene? Estimate the time it would take you to solve for the energy levels of naphthalene by hand using the brute force technique.

6. MOPAC (AM1) calculations yield a partial charge on the nitrogen atom in the ammonium ion of –0.094. What is the partial charge on each of the hydrogen atoms in the ammonium ion?

ChemActivity **17**

Molecular Orbitals and Energies for Diatomic Molecules in the Second Period.

(Are Molecular Orbitals Everywhere?)

Diatomic molecules of the second period, such as N_2 and O_2, are particularly important to us because of their prevalence and because they are often used as models for more complicated molecules.

Model 1: The Energy Levels of N_2 using MOPAC/AM1.

$:N \equiv N:$ (atoms 1 and 2)

Coordinate system on each atom.

$$\Psi_{\text{trial}} = c_1\, 2s_{N1} + c_2\, 2px_{N1} + c_3\, 2py_{N1} + c_4\, 2pz_{N1} + c_5\, 2s_{N2} + c_6\, 2px_{N2} + c_7\, 2py_{N2} + c_8\, 2pz_{N2} \quad (1)$$

Table 1. MOPAC Energy Levels for N_2.

Energy Level	Energy (eV)	Energy Level	Energy (eV)
1	-41.39	6	1.00
2	-21.43	7	1.00
3	-16.19	8	6.03
4	-16.19		
5	-14.32		

Table 2. MOPAC Wave functions for N_2.

	Coefficients for each MO (Ψ_i)				
Atomic Orbital	Ψ_1	Ψ_2	Ψ_3	Ψ_4	Ψ_5
2sN1	0.6210	-0.6497	0.0000	0.0000	-0.3382
2pxN1	0.3382	0.2792	0.0000	0.0000	0.6210
2pyN1	0.0000	0.0000	0.7071	0.0000	0.0000
2pzN1	0.0000	0.0000	0.0000	0.7071	0.0000
2sN2	0.6210	0.6497	0.0000	0.0000	-0.3382
2pxN2	-0.3382	0.2792	0.0000	0.0000	-0.6210
2pyN2	0.0000	0.0000	0.7071	0.0000	0.0000
2pzN2	0.0000	0.0000	0.0000	0.7071	0.0000

	Coefficients for each MO (Ψ_i)		
Atomic Orbital	Ψ_6	Ψ_7	Ψ_8
2sN1	0.0000	0.0000	-0.2792
2pxN1	0.0000	0.0000	-0.6497
2pyN1	0.5000	0.5000	0.0000
2pzN1	-0.5000	0.5000	0.0000
2sN2	0.0000	0.0000	0.2792
2pxN2	0.0000	0.0000	-0.6497
2pyN2	-0.5000	-0.5000	0.0000
2pzN2	0.5000	-0.5000	0.0000

Critical Thinking Questions

1. MO1 is drawn below.

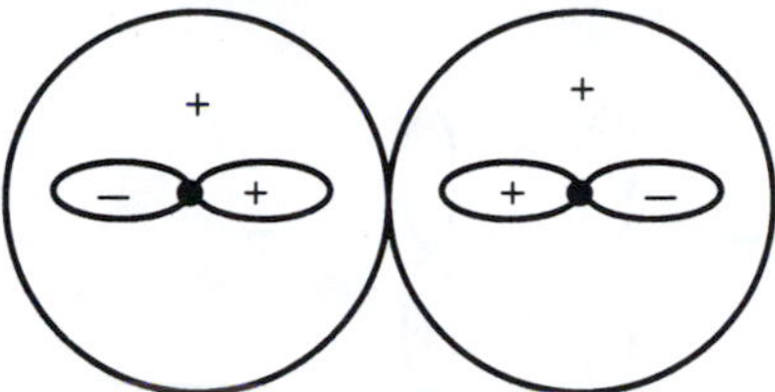

Make note of the coordinate system used in Model 1 and rationalize the signs shown in the figure for the $2px$N1 and $2px$N2 contributions.

2. Make a sketch of MO2 below.

3. The molecular orbital energy level diagram for N_2 is shown in Figure 1 (next page). For simplicity, only the two AOs (one from each atom) that contribute the most to the MOs are shown. Each MO has a designation, for example σ_g, written next to the sketch of the orbital. Recall that only the valence orbitals are used in the trial function and place the correct number of electrons into the energy level diagram to generate the ground electronic state of the molecule.

Figure 1. The MO energy level diagram for N_2.

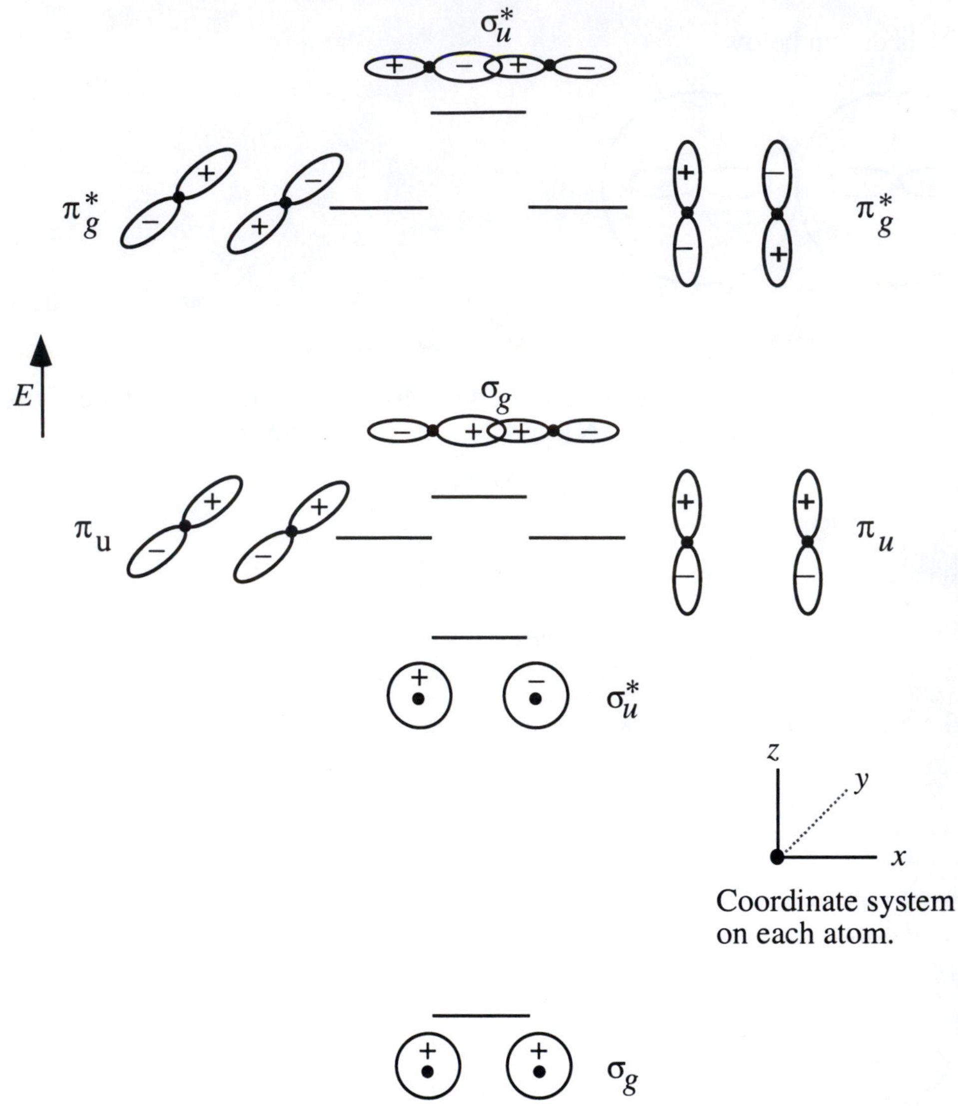

4. Examine the Lewis structure for N_2. Is the ground state of N_2 diamagnetic or paramagnetic according to the Lewis structure? Briefly explain.

5. According to MO theory (Model 1; Figure 1), is the ground state of N_2 diamagnetic or paramagnetic? Briefly explain.

6. The electron configuration for the 14 electrons in N_2 can be written:

$$1s^2\ 1s^2\ (\sigma_g)^2\ \ (\sigma_u^*)^2\ \ (\pi_u)^2\ \ (\pi_u)^2\ \ (\sigma_g)^2$$

On the axes below, sketch the photoelectron spectrum of N_2. The $1s^2$ electrons on each nitrogen atom are at 410 eV and are virtually unaffected by the bonding taking place in the second shell. The other ionization energies can be found in Table 1.

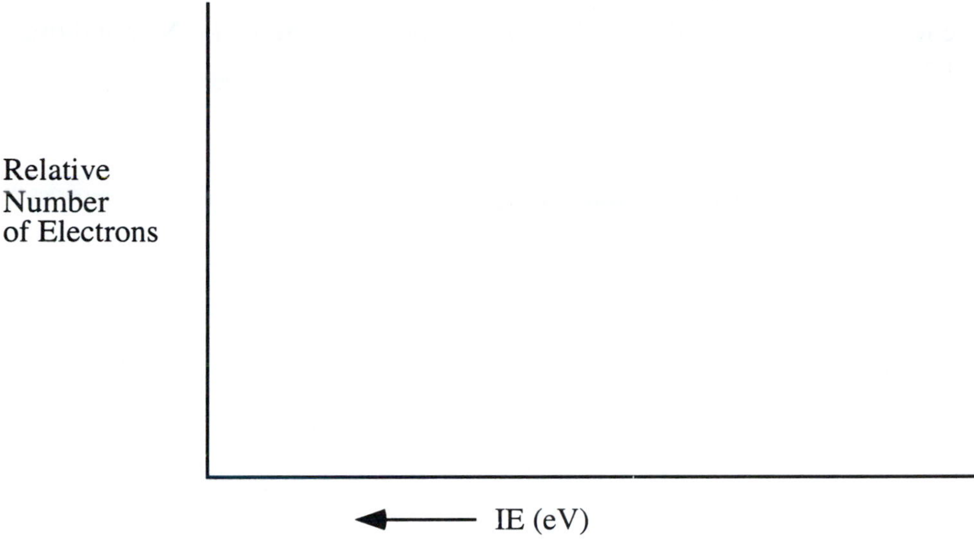

Information

Because a diatomic molecule contains only two atoms, any MO that increases the electron density between the nuclei is called a bonding orbital; any MO that has a node between the two atoms is called an antibonding orbital. For example, MO1 for N_2, σ_g, is a bonding orbital, and MO2 for N_2, σ_u^*, is an antibonding orbital.

Critical Thinking Questions

7. Classify each of the eight MOs of N_2 as bonding or antibonding.

8. What feature of the orbital designation indicates if the MO is antibonding?

Model 2: The Bond Order of the N–N Bond in N_2.

$$\text{Bond Order} = \frac{\text{number of bonding electrons} - \text{number of antibonding electrons}}{2} \qquad (2)$$

Critical Thinking Questions

9. The MOPAC/AM1 bond order for N_2 is 3.000. Determine the N_2 bond order by using Model 2.

10. Why is the factor "1/2" found in equation (2)?

Information

A MOPAC/AM1 treatment of any homonuclear diatomic molecule in the second period, say B_2 or F_2, would use exactly the same eight atomic orbitals as used for N_2. The energy levels for B_2 would be at somewhat higher energies, and MOs for F_2 would be somewhat lower energies, but the eight MOs would be the same. Thus the MO diagram shown in Figure 1 is used, without the specific energy levels, for homonuclear molecules, and *almost homonuclear* molecules (CO; CN^-, and so on) of the second period.

Figure 2. The MO Energy Level Diagram for Most Diatomic Molecules of the Second Period.

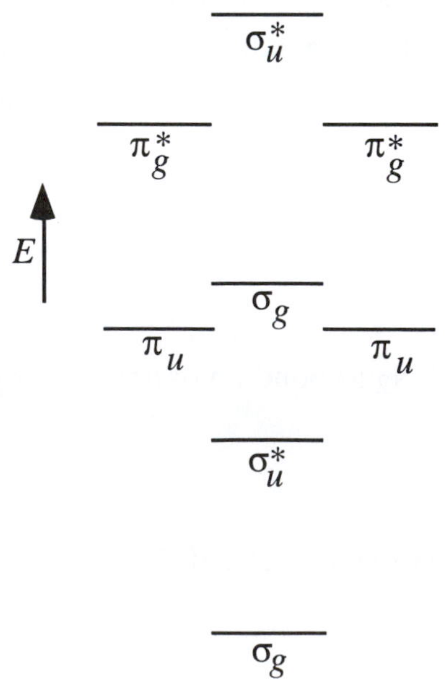

σ_u^*

π_g^* π_g^*

E

σ_g

π_u π_u

σ_u^*

σ_g

11. Write the Lewis structure for O_2. Is the ground state of O_2 diamagnetic or paramagnetic according to the Lewis structure? Briefly explain.

12. According to MO theory (use Figure 2), is the ground state of O_2 diamagnetic or paramagnetic? Briefly explain.

13. What experiment can you describe that would determine if O_2 is diamagnetic or paramagnetic?

14. Experiments prove that O_2 is paramagnetic (with two unpaired electrons). Comment on the reliability of Lewis structures to correctly predict the magnetic properties of molecules.

15. According to the Lewis structure, what is the bond order of the O–O bond in O_2?

16. According to MO theory (Figure 2), what is the bond order of the O–O bond in O_2?

Exercises

1. For the following molecules, use the diatomic MO energy level diagram in Figure 2 to a) determine the bond order, b) determine whether the molecule is diamagnetic or paramagnetic, and c) write the electron configuration of the molecule: B_2; F_2; CN; BF.

2. For each of the following pairs of molecules (or ions) use the diatomic MO energy level diagram in Figure 2 to determine which species has the stronger bond: O_2 and O_2^+; N_2 and CN^-; CO and CO^+; NO and NO^+.

Model 3: The Center of Inversion, i.

If a molecule can be brought to an equivalent arrangement of atoms (one that is indistinguishable from the original arrangement) by changing the coordinates (x, y, z) of every atom, where the origin of the coordinates lies at a point within the molecule, into ($-x$, $-y$, $-z$), then the point at which the origin lies is said to be a center of inversion, *i*.

For example, both N_2 and benzene have center of inversion.

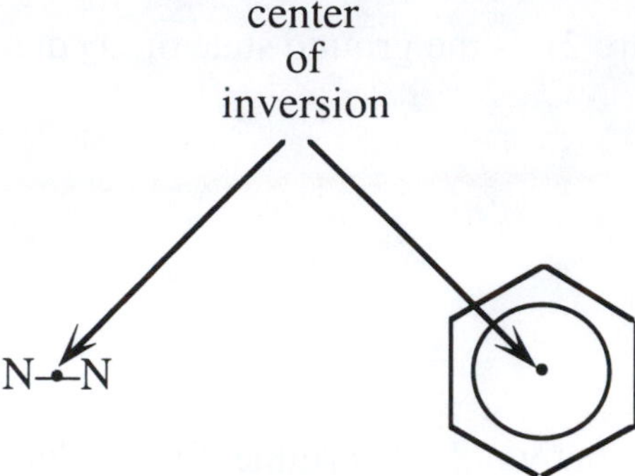

Whereas, CO and chlorobenzene do not have a center of inversion.

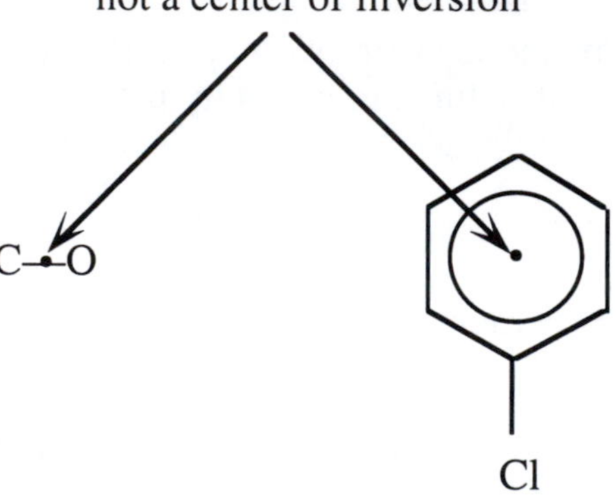

Critical Thinking Questions

17. Why does chlorobenzene not have a center of inversion?

18. Does ethylene have a center of inversion?

Exercise

3. Which of the following molecules has a center of inversion: O_2; CH_4; 1,4-dichlorobenzene; NO_3^-; H_2O?

Model 4: The Inversion Operation, $\hat{i}$.

The inversion operation, $\hat{i}$, takes a function and converts all (x, y, z) to $(-x, -y, -z)$.

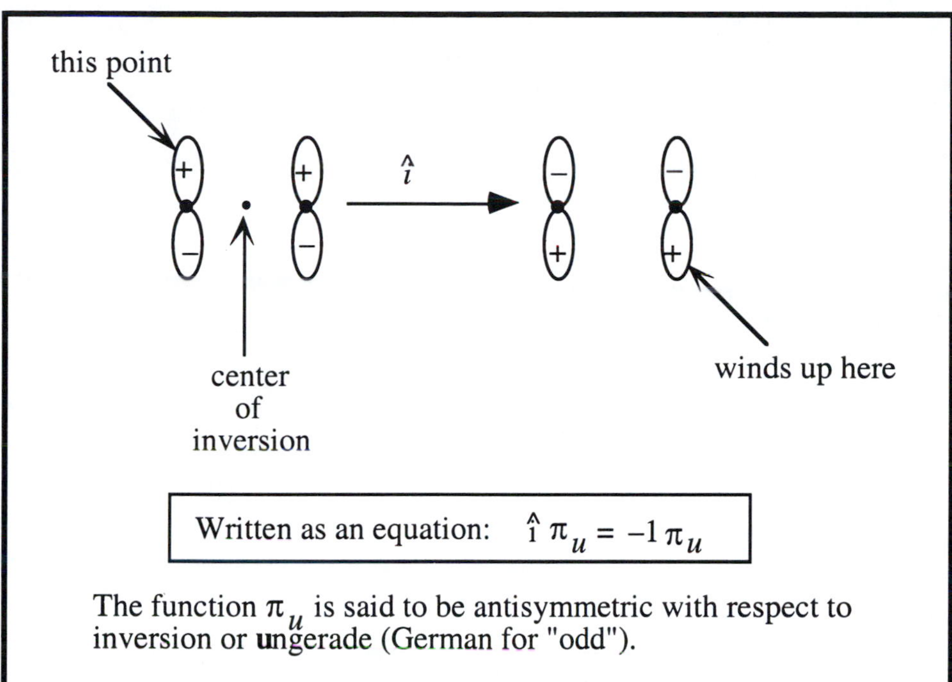

Written as an equation: $\hat{i}\,\pi_u = -1\,\pi_u$

The function π_u is said to be antisymmetric with respect to inversion or **un**gerade (German for "odd").

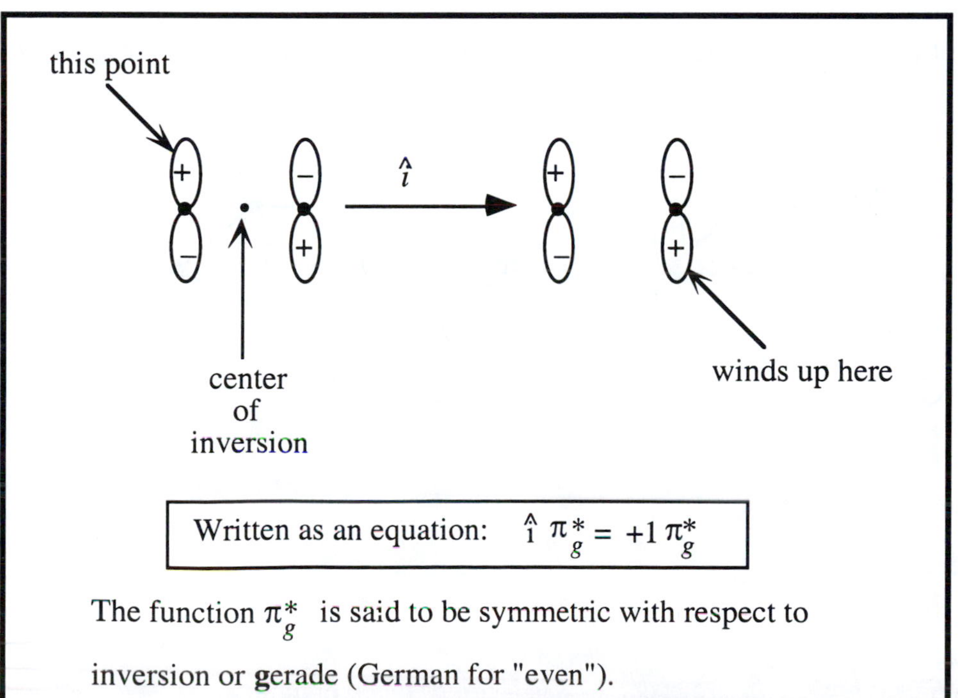

Written as an equation: $\hat{i}\,\pi_g^* = +1\,\pi_g^*$

The function π_g^* is said to be symmetric with respect to inversion or **g**erade (German for "even").

Critical Thinking Questions

19. Show that the highest energy MO for N_2, see Figure 1, is correctly labeled as ungerade.

20. If an orbital is ungerade, does that imply that it is antibonding?

Exercises

4. Is a 1*s* orbital on a hydrogen atom *u* or *g*?

5. Is a 2*p* orbital on a hydrogen atom *u* or *g*?

6. Is a 3*d* orbital on a hydrogen atom *u* or *g*?

7. Two of the MOs of benzene are shown below. Classify each of these MOs as *u* or *g*.

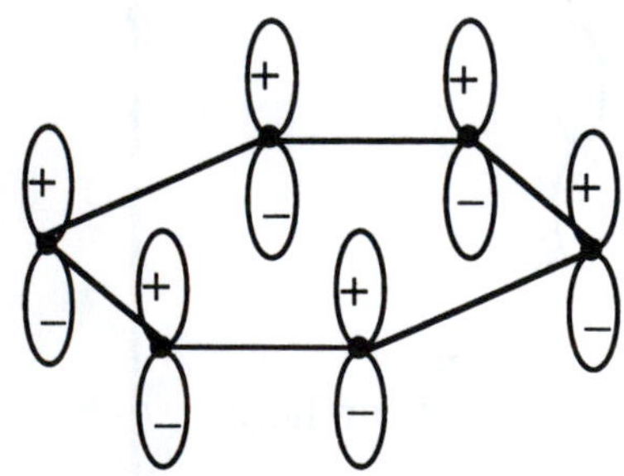

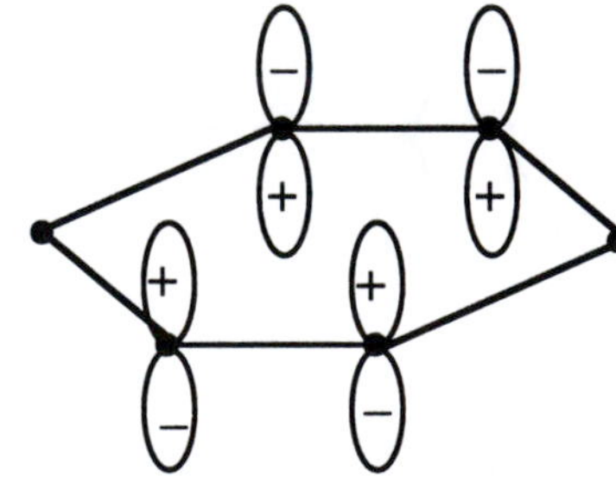

Model 5: The Orbital Angular Momentum of MOs.

Orbital Angular Momentum Quantum Number	One-Electron In An Atomic Orbital	Multielectron Atom	One-Electron In A Molecular Orbital	Multielectron Diatomic Molecule
0	*s*	*S*	σ	Σ
1	*p*	*P*	π	Π
2	*d*	*D*	δ	Δ
3	*f*	*F*		

For an electron in a σ molecular orbital:

$\ell = 0$ and $m = 0$

$s = 1/2$ and $m_s = 1/2$ or $-1/2$

For an electron in a π molecular orbital:

$\ell = 1$ and $m = +1$ or -1

$s = 1/2$ and $m_s = 1/2$ or $-1/2$

Homonuclear Energy Level Diagram for Most Molecules in the Second Period

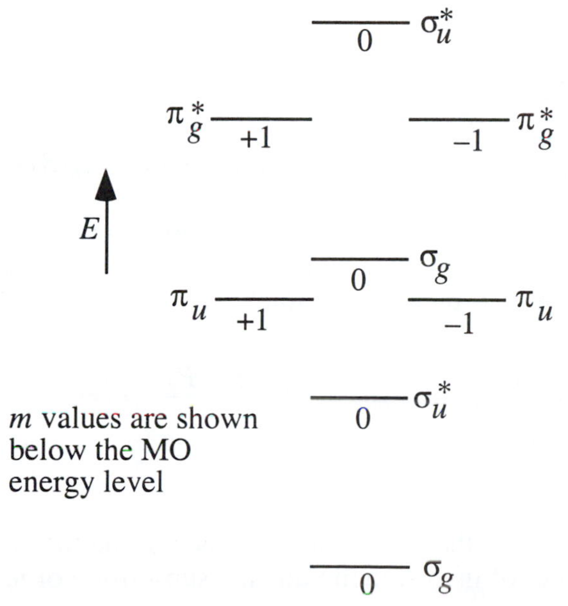

σ_u^* 0

π_g^* +1   −1 π_g^*

E

σ_g 0

π_u +1   −1 π_u

σ_u^* 0

m values are shown below the MO energy level

σ_g 0

Critical Thinking Questions

21. What is the orbital angular momentum quantum number, ℓ, for an electron described as σ_u* in N_2?

22. Use Model 5 to explain why the term symbol for the ground state of N_2 is $^1\Sigma$.

23. Use Model 5 to explain why the term symbol for the ground state of O_2^+ is $^2\Pi$.

Information

The complete term symbol for the ground electronic state of N_2 is $^1\Sigma_g$. The subscript *g* indicates that the total wave function for N_2 is gerade (even). Recall that the total wave function for the 10 valence electrons of N_2 is a product of the wave functions for the individual electrons:

$$\Psi_{total} = \Psi(1)\ \Psi(2)\ ...\ \Psi(10)$$

For the ground state of N_2:

$$\Psi_{total} = \sigma_g(1)\ \sigma_g(2)\ \sigma_u{}^*(3)\ \sigma_u{}^*(4)\ \pi_u(5)\ \pi_u(6)\ \pi_u(7)\ \pi_u(8)\ \sigma_g(9)\ \sigma_g(10)$$

Because an even function times an even function results in an even function, and an odd function times an odd function results in an even function, and an odd function times an even function results in an odd function, the total wave function for N_2 has symmetry *g*.

$$\Psi_{total} = g \times g \times u \times u \times u \times u \times u \times u \times g \times g = \Psi_g$$

Exercise

8. Find the term symbol for the ground state of each of the following molecules. If the molecule has a center of inversion include the subscript *g* or *u*.

   CO; F_2; O_2; NO^+

ChemActivity 18

Thermal Energies

(What is temperature?)

We have seen that the energy of a molecule can be separated into translational, rotational, vibrational, and electronic energies. Within each type of energy, the energy levels are quantized. Each energy state is described by a wave function which depends on a quantum number or a set of quantum numbers. For example, a particular molecule of dihydrogen, H_2, in a box could be described as follows:

translational:	$n_x = 4.12 \times 10^7$, $n_y = 8.22 \times 10^2$, $n_z = 4.37 \times 10^{11}$
rotational:	$J = 12$, $m = 5$
vibrational:	$\upsilon = 0$
electronic:	${}^1\Sigma_g$ or (σ_g^2) the electronic ground state

This molecule is in the ground electronic state. The two electrons are paired in a gerade bonding orbital. The molecule has no spin angular momentum. The molecule has no orbital angular momentum. The molecule is in the lowest possible vibrational energy state. The molecule has a modest amount of rotational energy. The molecule is moving fastest in the z direction and slowest in the y direction.

We now turn our attention to how a collection of molecules is distributed throughout the available energy states.

Model 1: A Hypothetical Molecule with only Two States at Different Energy Levels.

● = molecule

$T = 0$ K

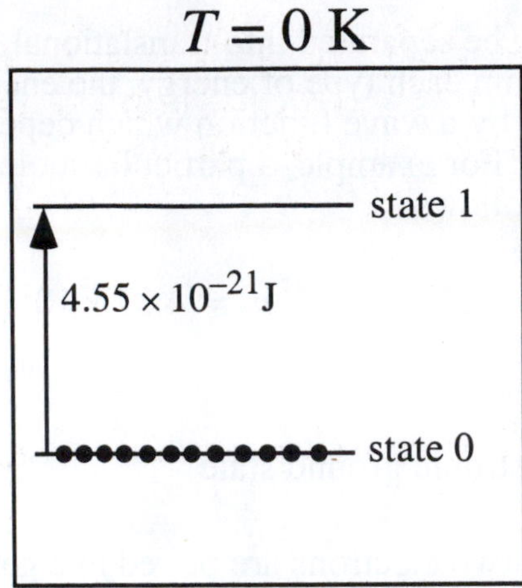

All of the molecules are in the lowest possible energy level at $T = 0$ K.

$T = 300$ K

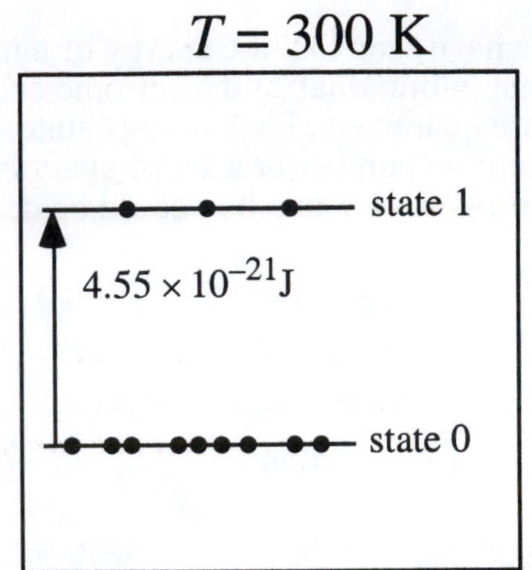

As the temperature increases, some of the molecules gain energy. One-fourth of the molecules are in the excited state at $T = 300$ K

$T = \infty$ K

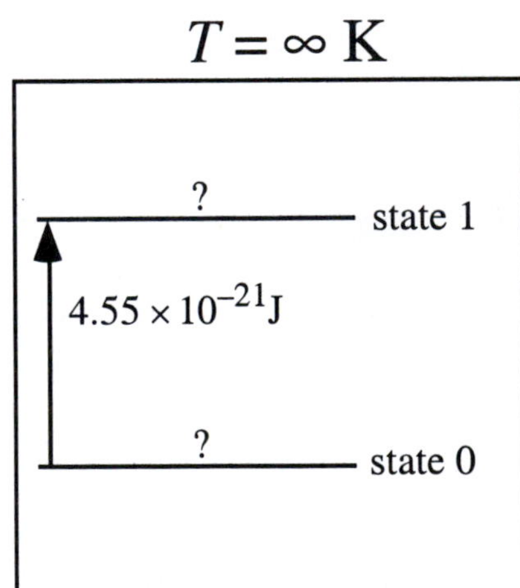

Information

The thermal energy is defined as the energy that the system has because it is at some temperature other than zero K.

Critical Thinking Questions

1. Why is state 1 not populated at $T = 0$ K?

2. Why does state 1 become populated at a temperature greater than $T = 0$ K?

3. Predict the number of molecules in state 0 and in state 1 at $T = \infty$ K.

4. What is the thermal energy for the twelve molecules in Model 1 at $T = 0$ K?

5. What is the thermal energy for the twelve molecules in Model 1 at $T = 300$ K?

6. What is the thermal energy per molecule in Model 1 at $T = 300$ K?

7. What is the thermal energy per mole in Model 1 at $T = 300$ K?

Model 2: The Boltzmann Distribution.

Ludwig Boltzmann developed an equation that yields the ratio of the number of molecules at two different energy levels, *i* and *j*, when the molecules are in thermal equilibrium. The equation is known as the Boltzmann distribution equation.

$$\frac{N_i}{N_j} = \frac{g_i}{g_j} e^{-(\varepsilon_i - \varepsilon_j)/kT} \tag{1}$$

where

ε_i and ε_j are the energies associated with the two energy levels
g_i and g_j are the degeneracies of the i and j energy levels, respectively
k is Boltzmann's constant, 1.380×10^{-23} J/K
T is the temperature in K

Critical Thinking Questions

8. Use Boltzmann's equation to verify that all of the molecules will be found in state 0 at $T = 0$ K (Model 1). [Hint: determine the ratio N_1/N_0.]

9. Use Boltzmann's equation to verify that three molecules of twelve will be found in state 1 at $T = 300$ K (Model 1).

10. Use Boltzmann's equation to determine the number of molecules in state 1 at $T = \infty$ K (Model 1).

11. Compare your prediction (CTQ 3) to the result obtained from the Boltzmann distribution equation.

Information

At infinite temperature there is sufficient energy available such that the molecules will simply populate all energy states without regard to energy. If there are 10 states, all will be equally occupied at $T = \infty$ K.

Exercises

1. For one mole of molecules, find the number of molecules in state 0 and state 1 at $T = 0$, 300, and ∞ K for the following system. Note that energy level 1 is two-fold degenerate.

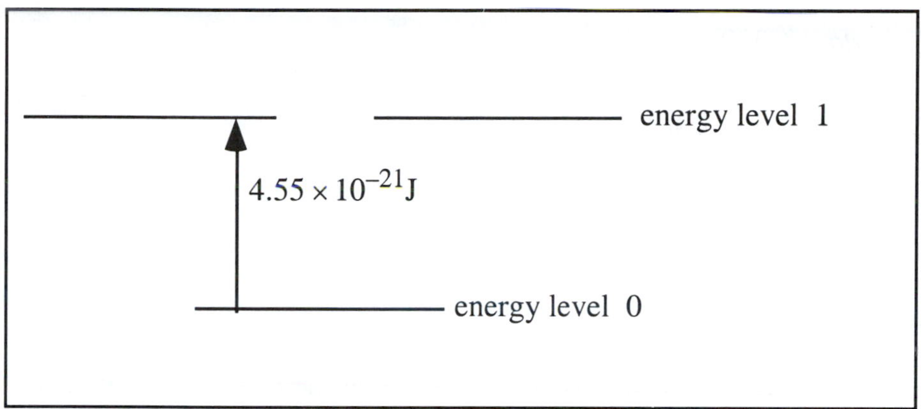

2. A system has three energy levels: ε_0, ε_1, and ε_2. The degeneracies of these energy levels are as follows: $g_0 = 2$; $g_1 = 1$; $g_2 = 9$. If one mole of molecules are placed into the system at $T = \infty$ and allowed to come to thermal equilibrium, how many molecules will be found in the ε_0 energy level?

Model 3: The Partition Function.

$$U_{\text{thermal}} = \frac{N}{q} kT^2 \frac{dq}{dT} \tag{2}$$

where:

N is the total number of molecules

$$q = \sum_{i=0}^{i_{max}} g_i \, e^{-(\varepsilon_i - \varepsilon_0)/kT} \tag{3}$$

q is called the **partition function**

Critical Thinking Questions

12. Show that the partition function for the system in Model 1 is

$$q = 1 + e^{-(330)/T}$$

13. Show that $\frac{dq}{dT} = \frac{330}{T^2} e^{-(330)/T}$

14. Evaluate q and dq/dT at 300 K. Use the equation for the thermal energy, equation (2) to calculate the thermal energy for 12 molecules at 300 K (Model 1). How does your value compare to the value calculated in CTQ 5?

Exercise

3. Find the thermal energy of one mole of molecules described by the system in Exercise 1 at 300 K. Explain in words why one mole of molecules described by the system in Exercise 1 has more thermal energy than one of molecules described by Model 1.

ChemActivity 19

Thermal Energies of Molecules

(Is it getting hot in here?)

We have seen that the distribution of molecules throughout the available state is determined by the temperature. Or, alternatively, a given distribution of molecules determines the temperature. The implication is that *temperature is a statistical phenomenon*. That is:

- If the distribution of gaseous molecules in a burning match and the distribution of molecules in a raging blast furnace are the same, the burning match and the blast furnace are at the same temperature.

- If one mole of N_2 molecules is steaming out of a nozzle and every molecule is moving with the same velocity (this would be hard to accomplish), the nitrogen stream has no temperature (strictly speaking) because there is no distribution of velocities. That is, at 100 K a collection of nitrogen molecules at thermal equilibrium should have a collection of velocities as given by the Boltzmann distribution; at 500 K the distribution would be somewhat different. If all of the molecules have the same velocity, this does not correspond to a Boltzmann distribution at any temperature. Temperature and the Boltzmann distribution are inextricably connected.

Concisely stated: the temperature determines the distribution and the distribution determines the temperature.

Now, we examine the nature of the distribution for translational, rotational, vibrational, and electronic energies.

Model 1: The Rotational Thermal Energy of Carbon Monoxide.

The rotational energy levels of a diatomic molecule are given by:

$$\varepsilon_J = J(J+1)\,\frac{h^2}{8\pi^2 I} \qquad J = 0, 1, 2, \ldots \qquad m = 0, \pm1, \pm2, \ldots, \pm J \tag{1}$$

The energy levels of carbon monoxide can be experimentally observed with microwave spectroscopy, and we find that

$$\varepsilon_J = J(J+1)\, 3.836 \times 10^{-23} \text{ Joules} \tag{2}$$

$$g_J = 2J + 1 \tag{3}$$

Recall that

$$q = \sum_{i=0}^{i_{max}} g_i\, e^{-(\varepsilon_i - \varepsilon_0)/kT}$$

$$q = 1 + 3\,e^{-\varepsilon_1/kT} + 5\,e^{-\varepsilon_2/kT} + 7\,e^{-\varepsilon_3/kT} + \ldots \tag{4}$$

$$\frac{dq}{dT} = 3\,\frac{\varepsilon_1}{kT^2}\,e^{-\varepsilon_1/kT} + 5\,\frac{\varepsilon_2}{kT^2}\,e^{-\varepsilon_2/kT} + 7\,\frac{\varepsilon_3}{kT^2}\,e^{-\varepsilon_3/kT} + \ldots \tag{5}$$

$$\frac{dq}{dT} = \frac{1}{kT^2}\left(3\,\varepsilon_1\,e^{-\varepsilon_1/kT} + 5\,\varepsilon_2\,e^{-\varepsilon_2/kT} + 7\,\varepsilon_3\,e^{-\varepsilon_3/kT} + \ldots\right) \tag{6}$$

$$U_{\text{thermal,rot}} = \frac{N}{q}\,kT^2\,\frac{dq}{dT}$$

$$= \frac{N}{\left(1 + 3e^{-\varepsilon_1/kT} + 5e^{-\varepsilon_2/kT} + 7e^{-\varepsilon_3/kT} + \ldots\right)}\left(3\varepsilon_1 e^{-\varepsilon_1/kT} + 5\varepsilon_2 e^{-\varepsilon_2/kT} + 7\varepsilon_3 e^{-\varepsilon_3/kT} + \ldots\right) \tag{7}$$

Table 1. Various Quantities for the Determination of q_{rot} and dq/dT for CO at 300 K.

J	g_J	ε_J	$g_J\, e^{-\varepsilon_J/kT}$	$\varepsilon_J\, g_J\, e^{-\varepsilon_J/kT}$
0	1	0.00	1.00E+00	
1	3	7.67 E-23	2.94E+00	2.26E-22
2	5	2.30E-22	4.73E+00	1.09E-21
3	7	4.60E-22	6.26E+00	2.88E-21
4	9	7.67E-22	7.48E+00	5.74E-21
5	11	1.15E-21	8.33E+00	9.59E-21
6	13	1.61E-21	8.81E+00	1.42E-20
7	15	2.15E-21	8.93E+00	1.92E-20
8	17	2.76E-21	8.72E+00	2.41E-20
9	19	3.45E-21	8.25E+00	2.85E-20
10	21	4.22E-21	7.58E+00	3.20E-20
11	23	5.06E-21	6.77E+00	3.43E-20
12	25	5.98E-21	5.89E+00	3.53E-20
13	27	6.98E-21	5.00E+00	3.49E-20
14	29	8.06E-21	4.14E+00	3.34E-20
15	31	9.21E-21	3.35E+00	3.09E-20
16	33	1.04E-20	2.65E+00	2.77E-20
17	35	1.17E-20	2.05E+00	2.41E-20
18	37	1.31E-20	1.56E+00	2.04E-20
19	39	1.46E-20	1.15E+00	1.68E-20
20	41	1.61E-20	8.37E-01	1.35E-20
21	43	1.77E-20	5.95E-01	1.05E-20
22	45	1.94E-20	4.14E-01	8.04E-21
23	47	2.12E-20	2.82E-01	5.98E-21
24	49	2.30E-20	1.89E-01	4.34E-21
25	51	2.49E-20	1.24E-01	3.08E-21
26	53	2.69E-20	7.93E-02	2.14E-21
27	55	2.90E-20	4.99E-02	1.45E-21
28	57	3.11E-20	3.08E-02	9.59E-22
29	59	3.34E-20	1.86E-02	6.21E-22
30	61	3.57E-20	1.10E-02	3.94E-22
31	63	3.81E-20	6.42E-03	2.44E-22
32	65	4.05E-20	3.66E-03	1.48E-22
33	67	4.30E-20	2.05E-03	8.81E-23
34	69	4.56E-20	1.12E-03	5.12E-23
35	71	4.83E-20	6.04E-04	2.92E-23
36	73	5.11E-20	3.19E-04	1.63E-23
37	75	5.39E-20	1.65E-04	8.89E-24
38	77	5.68E-20	8.37E-05	4.76E-24
39	79	5.98E-20	4.17E-05	2.50E-24
40	81	6.29E-20	2.04E-05	1.28E-24
41	83	6.61E-20	9.77E-06	6.45E-25
42	85	6.93E-20	4.59E-06	3.18E-25
43	87	7.26E-20	2.12E-06	1.54E-25
44	89	7.60E-20	9.59E-07	7.28E-26
45	91	7.94E-20	4.26E-07	3.38E-26
46	93	8.29E-20	1.86E-07	1.54E-26
47	95	8.65E-20	7.93E-08	6.87E-27
48	97	9.02E-20	3.33E-08	3.00E-27
49	99	9.40E-20	1.37E-08	1.29E-27
50	101	9.78E-20	5.53E-09	5.41E-28
Sum			1.08E+02	4.47E-19

Critical Thinking Questions

1. Verify that equation (3) is correct.

2. Verify that all of the entries for $J = 1$ in Table 1 are correct.

3. Use equation (7) and Table 1 to show that $U_{thermal,rot}$ = 2.49 kJ/mole for CO at 300 K.

4. The Boltzmann distribution for the ratio of the number of molecules in energy level J to the number of molecules in $J = 0$ is:

$$\frac{N_i}{N_j} = \frac{g_i}{g_j}\, e^{-(\varepsilon_i - \varepsilon_j)/kT} = \frac{N_J}{N_o} = g_J\, e^{-\varepsilon_J/kT}$$

This ratio is identical to column four in Table 1. What is the ratio N_{10}/N_0 for CO at 300 K?

5. According to Table 1, which rotational energy level is most highly populated? What is the value of N_J/N_0 for this energy level?

6. What is the value of N_{50}/N_0 ?

7. Which molecules are more responsible for the thermal energy of 2.49 kJ/mole for CO at 300 K—the molecules in energy level $J = 50$ or the molecules in energy level $J = 10$?

8. Why was the summation stopped at $J = 50$? Could the summation have been stopped sooner? Explain.

Exercises

1. For CO, use the Boltzmann distribution equation along with equations (1), (2), and (3) to calculate the ratio of the number of molecules in rotational energy ε_1 to the number of molecules in ε_0 (that is, N_1/N_0) at 300 K. Calculate N_{50}/N_0 at 300 K.

2. It can be shown that for all linear molecules with unlike ends (HCl, HCN):

$$q_{\text{rot}} = \frac{8\pi^2 Ik}{h^2} \times T$$

   a) Use this equation to show that for one mole of molecules with unlike ends:

$$U_{\text{rot, thermal}} = RT$$

   b) Use the result from part a) to calculate the rotational thermal energy of one mole of CO at 300 K. How does this result compare to your result from CTQ 3?

Model 2: The Vibrational Thermal Energy of Carbon Monoxide.

The vibrational energy levels of a diatomic molecule are given by:

$$\varepsilon_\upsilon = (\upsilon + \frac{1}{2}) \frac{h}{2\pi} \sqrt{\frac{k}{\mu}} \qquad \upsilon = 0, 1, 2, ... \tag{8}$$

The energy levels of carbon monoxide can be experimentally observed with infrared spectroscopy, and we find that

$$\varepsilon_\upsilon = (\upsilon + \frac{1}{2})\, 4.257 \times 10^{-20} \text{ Joules} \tag{9}$$

$$g_\upsilon = 1 \tag{10}$$

$$q = 1 + e^{-(\varepsilon_1 - \varepsilon_0)/kT} + e^{-(\varepsilon_2 - \varepsilon_0)/kT} + \ldots \tag{11}$$

$$\frac{dq}{dT} = \frac{(\varepsilon_1 - \varepsilon_0)}{kT^2} e^{-(\varepsilon_1 - \varepsilon_0)/kT} + \frac{(\varepsilon_2 - \varepsilon_0)}{kT^2} e^{-(\varepsilon_2 - \varepsilon_0)/kT} + \ldots \tag{12}$$

$$U_{\text{thermal,vib}} = \frac{N}{q} kT^2 \frac{dq}{dT}$$

$$= \frac{N}{\left(1 + e^{-(\varepsilon_1 - \varepsilon_0)/kT} + e^{-(\varepsilon_2 - \varepsilon_0)/kT} + \ldots\right)} \left((\varepsilon_1 - \varepsilon_0) e^{-(\varepsilon_1 - \varepsilon_0)/kT} + (\varepsilon_2 - \varepsilon_0) e^{-(\varepsilon_2 - \varepsilon_0)/kT} + \ldots\right) \tag{13}$$

Table 2. Various Quantities for the Determination of q_{vib} and dq/dT for CO at 300 K.

υ	ε_υ	$\varepsilon_\upsilon - \varepsilon_0$	$e^{-(\varepsilon_\upsilon - \varepsilon_0)/kT}$	$(\varepsilon_\upsilon - \varepsilon_0)\, e^{-(\varepsilon_\upsilon - \varepsilon_0)/kT}$
0	2.13E-20	0.00E+00	1.00E+00	
1	6.39E-20	4.26E-20	3.42E-05	1.46E-24
2	1.06E-19	8.51E-20	1.17E-09	9.97E-29
Sum			1.00E+00	1.46E-24

Critical Thinking Questions

9. Verify that all of the entries for $\upsilon = 1$ in Table 2 are correct.

10. Use equation (13) and Table 2 to show that at 300 K, $U_{thermal,vib}$ = 0.879 J/mole for CO.

11. According to Table 2, which vibrational energy level is most populated?

12. About 50 energy levels were used to determine the rotational thermal energy. Only three energy levels were used to determine the vibrational thermal energy. Why?

13. The rotational thermal energy for CO is greater than the vibrational thermal energy. Why? [Hint: examine the Boltzmann distribution with regard to the value of $\varepsilon_i - \varepsilon_o$.]

Exercises

3. CO has a very strong bond (a triple bond), the separation between vibrational energy levels is quite large, and the thermal vibrational energy is quite small (0.877 J/mole). Find the thermal vibrational energy at 300 K for a diatomic molecule with a much weaker bond (a single bond), Cl_2. The vibrational energy levels of Cl_2 are given by:

$$\varepsilon_\upsilon = (\upsilon + \frac{1}{2})\ 1.11 \times 10^{-20} \text{ Joules}$$

4. Use the Boltzmann distribution equation along with equations (8), (9), and (10) to calculate N_1/N_0 (vibrational energy levels) for CO at 300 K. Is your result in agreement with the low thermal vibrational energy for CO at 300 K? Why or why not?

Model 3: The Electronic Thermal Energy of Carbon Monoxide.

The lowest energy singlet state of the CO molecule is 1.286×10^{-18} J above the ground electronic state.

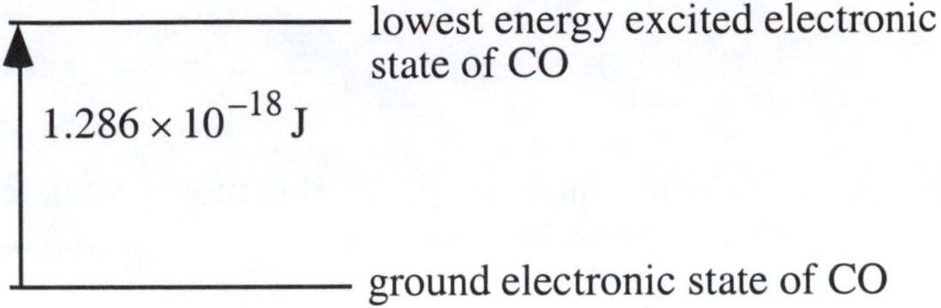

Critical Thinking Questions

14. For CO, is the energy difference between the ground electronic state and the excited electronic state larger or smaller than the energy difference between the $\upsilon = 0$ vibrational state and the $\upsilon = 1$ excited vibrational state?

15. Assume the ground and first excited electronic energy states of CO are singly degenerate. Do you expect the thermal electronic energy to be higher or lower than the thermal vibrational energy?

Exercises

5. For one mole of CO molecules, use the Boltzmann distribution equation to calculate the number of molecules in the ground electronic state and the number of molecules in the first excited electronic state of CO at 300 K. [Hint: $N_{\text{ground}} + N_{ex} = N_{A}$.] Calculate the thermal electronic energy of CO at 300 K.

6. Comment of the following statement:

   At normal temperatures the electronic thermal energy of any molecule is essentially zero.

Model 4: The Translational Thermal Energy of Carbon Monoxide.

The translational energy levels of a molecule in a container (particle-in-a-box) are given by:

$$\varepsilon = \varepsilon(\mathrm{x}) + \varepsilon(\mathrm{y}) + \varepsilon(\mathrm{z}) = \frac{n_x^2 h^2}{8ma^2} + \frac{n_y^2 h^2}{8mb^2} + \frac{n_z^2 h^2}{8mc^2} \qquad (14)$$

where a, b, and c, are the dimensions of the container, and each $n_i = 1, 2, 3, \ldots$.

It can be shown that the translational partition function is:

$$q_{\text{trans}} = \left(\frac{2\pi mk}{h^2}\right)^{3/2} V\,T^{3/2} \qquad (15)$$

where m is the mass of one molecule and V is the volume of the container.

Critical Thinking Questions

16. Use the translational partition function to show that $U_{\text{trans, thermal}} = \frac{3}{2}RT$ for one mole of gas. Recall that:

$$U_{\text{thermal}} = \frac{N}{q}\,kT^2\,\frac{dq}{dT}$$

17. At 300 K, what is $U_{\text{trans, thermal}}$ for one mole of CO(g)?

18. At 300 K, what is $U_{\text{rot, thermal}}$ for one mole of CO(g)?

19. At 300 K, what is $U_{\text{vib, thermal}}$ for one mole of CO(g)?

20. At 300 K, what is $U_{\text{elec, thermal}}$ for one mole of CO(g)?

21. At 300 K, what is $U_{\text{total, thermal}}$ for one mole of CO(g)?

22. At 300 K, which contributes more to the thermal energy of carbon monoxide—the translational thermal energy, the rotational thermal energy, the vibrational thermal energy or the electronic thermal energy?

Exercises

7. Estimate the total thermal energy (in kJ/mole) of Ne gas at 300.0 K. Estimate the total thermal energy of Ne gas at 301.0 K. The heat capacity of a substance is thermal energy change per mole per K. Estimate the heat capacity of Ne at 300 K, in J/(K mole) from your calculations above. Compare your value to the experimental heat capacity of Ne at 300 K, 12.47 J/K mole.

8. The translational thermal energy of a gas is given by $U_{\text{trans, thermal}} = \frac{3}{2} RT$ (CTQ 15). If the heat capacity is defined as dU/dT, what is the expression for the heat capacity of a gas with translational thermal energy only? Calculate a heat capacity for Ne based on this expression. Compare your value to the values in Exercise 7.

ChemActivity 20

Selection Rules

(Do you always obey the rules?)

We have seen that the total energy of atoms and molecules can be separated into the following components:

- translational energy (atoms and molecules)
- rotational energy (molecules)
- vibrational energy (molecules)
- electronic energy (atoms and molecules)

Now we turn our attention to how the energy of an atom or a molecule can be changed. Fundamentally there are two ways :

- collisions between two or more molecules, atoms, ions, or other particles
- absorption or emission of a photon

Collisions between molecules can be violent, and one can conceive of almost any transition taking place in a collision. Molecules gain translational energy (and lose translational energy) by collisions with other molecules; molecules do not increase or decrease their translational energy by absorption of electromagnetic radiation. For example, molecule A moving with a certain speed might collide with a molecule B moving at a different speed—as a result of a collision, molecule A might move faster and molecule B might move slower (there must be a conservation of energy). Or, molecule B might be in rotational state $J = 12$ before the collision and in rotational state $J = 5$ after the collision—in which case, molecule A might be moving faster (having gained the difference in energy between rotational states 12 and 5. These sorts of transitions are both common and important. In this activity, however, we turn our attention to transitions that take place with the absorption or emission of a photon.

Model 1: The Planck-Einstein Relationship

$$E_{\text{photon}} = h\nu = \frac{hc}{\lambda}$$

Region	Wavelength Range
radiowave	3 km – 30 cm
microwave	30 cm – 1 mm
infrared (IR)	1 mm – 800 nm
visible (VIS)	800 nm – 400 nm
ultraviolet (UV)	400 nm – 10 nm
X-ray	10 nm – 0.1 nm
gamma ray	< 0.1 nm

Critical Thinking Questions

1. What is the energy, in joules, of a typical x-ray photon?

2. What is the energy, in joules, of a typical microwave photon?

Exercises

1. What is the ratio of the energy of a typical x-ray photon to the energy of a typical microwave photon?

2. Rank the following photons in order of increasing energy: UV, microwave, x-ray, visible, radiowave, IR, gamma ray.

3. Complete the following table:

Energy (J)	Wavelength (m)	Frequency (s^{-1})	Region of Spectrum
		3.00×10^{9}	
	0.500×10^{-6}		
9.94×10^{-19}			
		2.00×10^{14}	
2.00×10^{-28}			
	1.00×10^{-9}		

Information

When a photon is emitted or absorbed by a molecule (or atom), the molecule changes from one energy state (designated by a set of quantum numbers) to another energy state (designated by a new set of quantum numbers). The difference in energy between the two states is equal to the energy of the photon. This is known as the Bohr frequency rule:

$$E_{photon} = h\nu = \frac{hc}{\lambda} = \varepsilon_{higher} - \varepsilon_{lower}$$

where ε_{higher} and ε_{lower} are the energy values for the for the higher and lower energy states of the molecule or atom.

Model 2: Typical Energy Level Spacings for the Various Types of Molecular Motion.

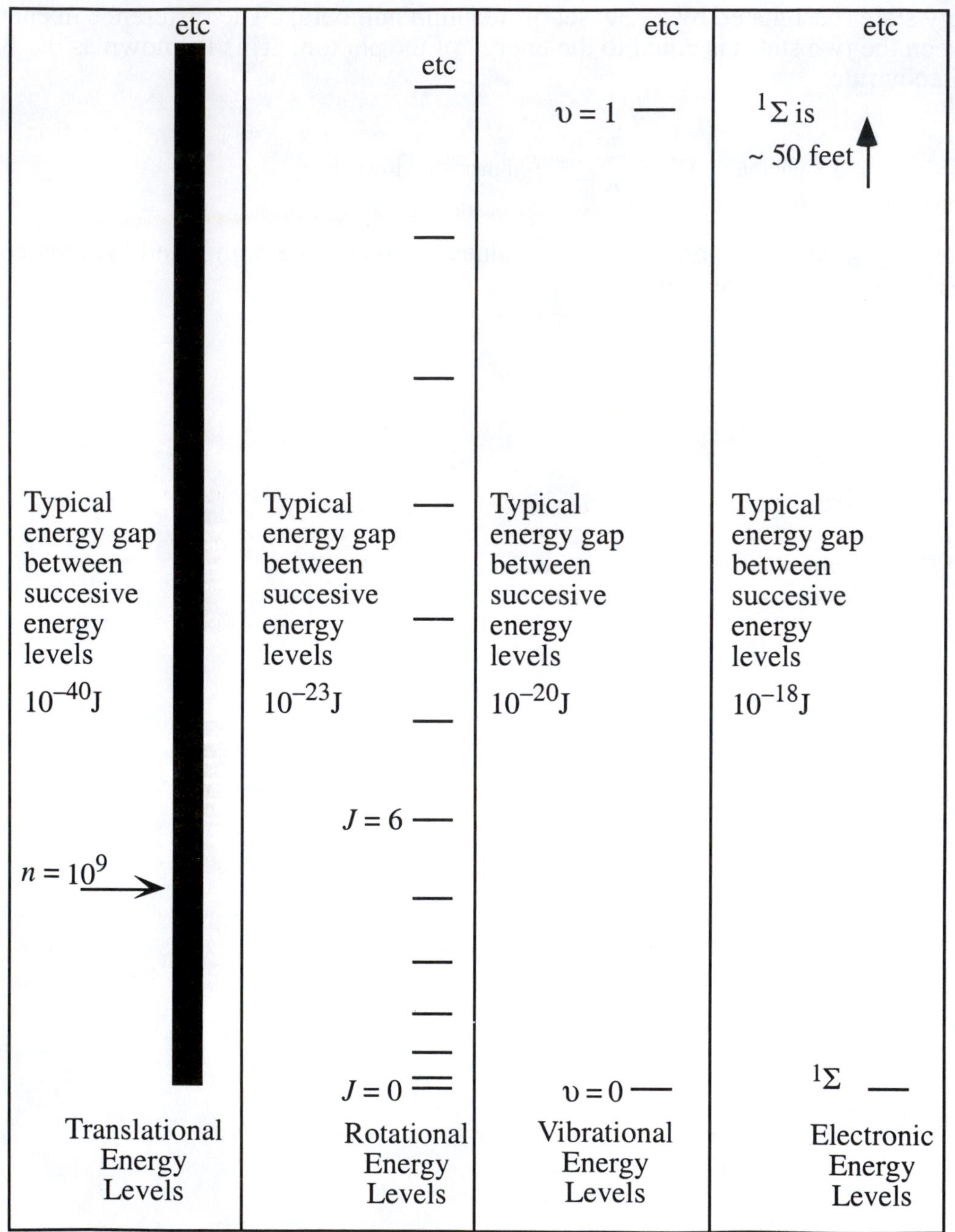

Note: Degeneracies of various energy levels are not shown.

Critical Thinking Questions

3. What is the name of the quantum mechanics model used for the calculation of translational energies?

4. Use the information in Model 2 to calculate the wavelength of the radiation commensurate with the energy level spacing for translational energies.

5. To experimentally measure a wavelength it is necessary to have an antenna or some device that has a length close to the wavelength being measured. Comment on the likelihood of constructing such an antenna for the wavelength calculated in CTQ 4.

6. What is the name of the quantum mechanics model used for the calculation of rotational energies for diatomic molecules?

7. Use the information in Model 2 to calculate the wavelength of the radiation commensurate with the energy level spacing for rotational energies. In what region of the electromagnetic spectrum is this wavelength?

8. What is the name of the quantum mechanics model used for the calculation of vibrational energies for diatomic molecules?

9. Use the information in Model 2 to calculate the wavelength of the radiation commensurate with the energy level spacing for vibrational energies. In what region of the electromagnetic spectrum is this wavelength?

10. What is the name of a quantum mechanics model used for the calculation of electronic energies for molecules?

11. Use the information in Model 2 to calculate the wavelength of the radiation commensurate with the energy level spacing for electronic energies. In what region of the electromagnetic spectrum is this wavelength?

Exercises

4. A single bond energy is typically about 300 kJ/mole. What is the minimum energy of a photon require to break a typical single bond (in one molecule)? What is the maximum wavelength of a photon capable of breaking a typical single bond? In what region of the electromagnetic spectrum is this photon? Which regions of the electromagnetic spectrum have sufficient energy to break a typical single bond? Which regions of the electromagnetic spectrum do not have sufficient energy to break a typical single bond?

5. Explain why UV radiation (not infrared or microwave) must be used in photochemical reactions, such as the Diels-Alder reaction shown below:

$$2\ H_2C{=}CH_2 \xrightarrow{UV} \begin{array}{c} CH_2{-}CH_2 \\ | \quad\quad | \\ CH_2{-}CH_2 \end{array}$$

6. Why is ultraviolet (and to some extent visible) electromagnetic radiation generally more harmful, photon for photon, to biological organisms than radiowaves, microwaves, and infrared?

Information

We have developed a number of models and wave functions to describe various **stationary states** or **time-independent states**, of atoms and molecules. Absorption and emission of electromagnetic radiation, however, is a **time-dependent process**. In order to absorb a photon, a molecule must have some aspect or feature that permits coupling with the oscillating electric (or magnetic) field of the photon.

Critical Thinking Questions

12. In carbon monoxide, which atom has the partial positive charge?

13. In carbon monoxide, which atom has the partial negative charge?

14. If a CO molecule is placed between two oppositely charged plates, why does the carbon atom tend to orient toward the negative plate and the oxygen atom toward the positive plate?

Model 3: The Effect of an Oscillating Electric Field on a CO molecule and on a N_2 Molecule.

Consider a carbon monoxide molecule suspended between two metal plates. The charge on the metal plates varies sinusoidally with time.

– | C≡O | + | C≡O | + | C≡O | –

(a)   (b)   (c)

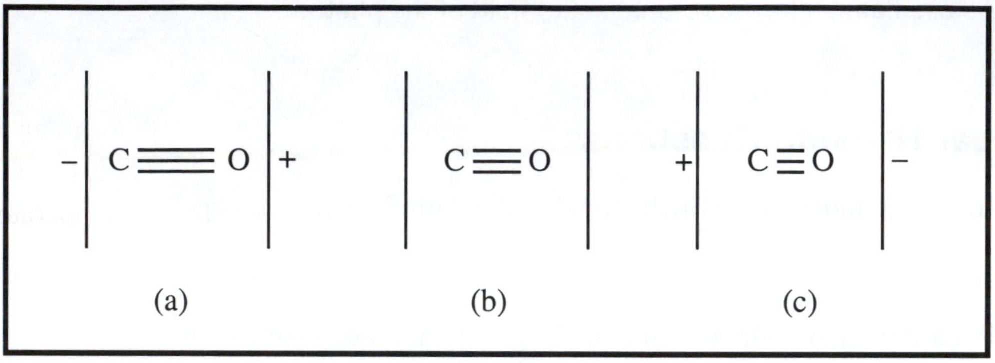

Consider a dinitrogen molecule suspended between two metal plates. The charge on the metal plates varies sinusoidally with time.

– | N≡N | + | N≡N | + | N≡N | –

(a)   (b)   (c)

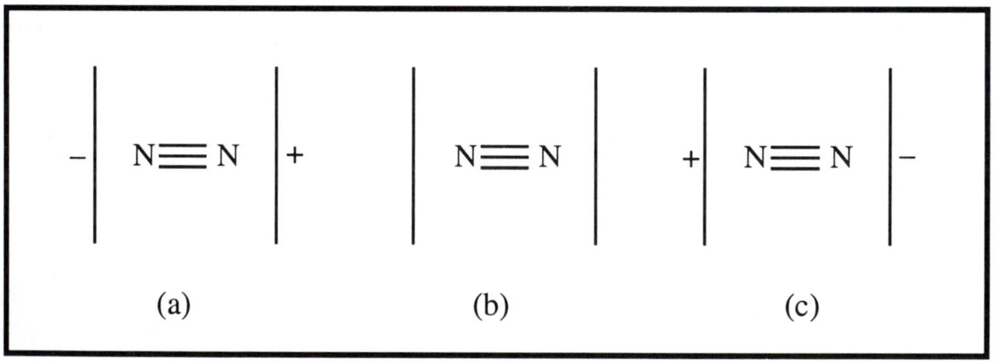

15. When the plates are charged as in (a), why don't the nitrogen atoms move toward the plates?

16. Why does the CO molecule move in response to an oscillating electric field, whereas the N_2 molecule does not?

Information

It can be shown that the probability of a transition between two quantum states described by Ψ_a and Ψ_b is proportional to:

- the intensity of the radiation of the proper wavelength,
- the number of molecules in the absorbing (or emitting) state,
- the square of the **transition dipole moment**. The transition dipole moment is an integral that contains the wave function (description) of the ground state (Ψ_a), the wave function of the excited state (Ψ_b), and the dipole moment operator, $\hat{u}$ (Model 3 demonstrated that the molecule must have some polarity to interact with the photon).

$$\text{transition dipole moment} = \int \Psi_a^* \hat{u} \Psi_b d\tau \qquad (1)$$

where $\hat{u}$ is the dipole moment operator.

As might be expected, evaluation of the transition dipole moment is not a trivial matter. In many cases, however, it is possible to determine if the transition dipole moment is zero or non-zero. These cases are summarized below:

- If $\int \Psi_a^* \hat{u} \Psi_b d\tau$ does not equal zero, the transition is predicted to occur (assuming, of course, that there are molecules in the absorbing or emitting state). The transition is said to be an **allowed transition**.
- If $\int \Psi_a^* \hat{u} \Psi_b d\tau$ does equal zero, the transition is not predicted to occur. The transition is said to be a **forbidden transition**.

A rule that relates which transitions are allowed is called a **selection rule**. The difference between allowed and forbidden transitions is not quite as stark as the terminology suggests. Forbidden transition do occur—but the intensity of a *forbidden* transition is 10^3 to 10^6 less than the intensity of the analogous *allowed* transition.

It is obvious that a photon can couple with a molecule that has a permanent dipole moment. However, the photon can also couple with a temporary or oscillating dipole moment. For example, we have seen that molecules are always vibrating (even at absolute zero). Thus, a molecule such as CO_2, which does not have a permanent dipole moment, will always have an oscillating dipole moment.

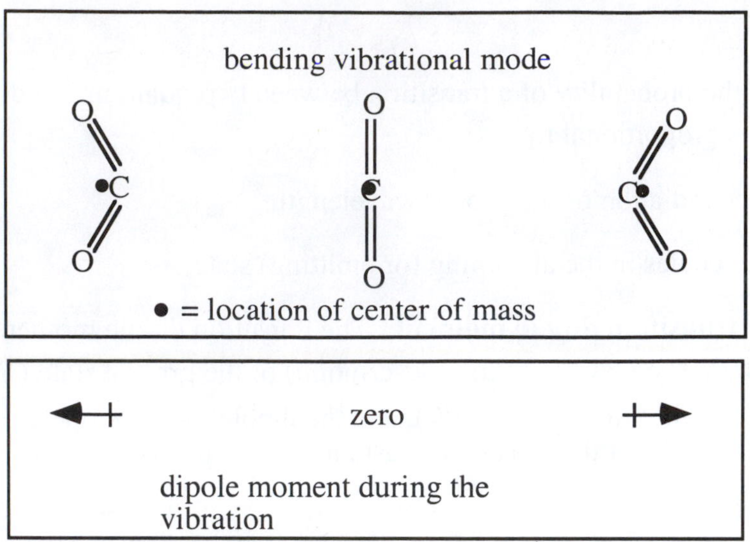

Critical Thinking Questions

17. Why is the probability of a transition proportional to the number of molecules in the absorbing (or emitting) state?

Exercises

7. Sketch the dipole moment vector of HF when $r = r_e$, when $r > r_e$, and when $r < r_e$.

8. Recall that in a vibrational mode the position of the center of mass does not change. Use a dot to represent the center of mass, and place the dot at the appropriate locations for the symmetrical stretch of SO_2, shown below. Then, draw arrows (see the bending vibrational mode of CO_2, above) to represent the dipole moment during the vibration.

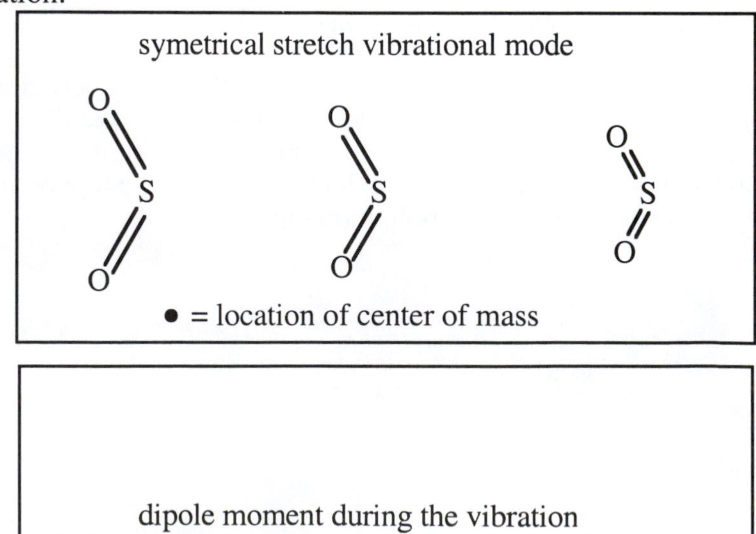

ChemActivity **21**

Rotational and Vibrational Spectra of Molecules

(Can you spin and pulsate at the same time?)

As mentioned previously, molecules do not increase their translational energy by absorption of electromagnetic radiation. Molecules gain translational energy (and lose translational energy) by collisions with other molecules. However, molecules do gain or lose rotational and vibrational energy by absorption (or emission) of electromagnetic radiation. In this activity we examine questions such as:

- How does a molecule in a particular rotational energy state, with quantum numbers J and m, undergo a transition that takes the molecule to a higher or lower rotational energy state, with quantum numbers J' and m'?
- How does a molecule in a particular vibrational energy state, with quantum number υ undergo a transition that takes the molecule to a higher or lower vibrational energy state?

Model 1: Rotational Selection Rules for Diatomic Molecules.

1. The molecule must have a permanent dipole moment. Homonuclear molecules like N_2 and very symmetric molecules like CO_2, CH_4, and SF_6 do not produce rotational spectra.

2. The transition must occur between adjacent rotational levels. For diatomic molecules, the selection rule is that ΔJ must equal ± 1.

Recall for the rigid rotor model (diatomic molecules):

$$\varepsilon_J = \frac{h^2}{8\pi^2 I_e} J(J+1) = B_e J(J+1) \qquad J = 0, 1, 2, \ldots \tag{1}$$

where r_e is the equilibrium internuclear distance

I_e is the equilibrium moment of inertia, $I_e = \mu r_e^2$

$$B_e = \frac{h^2}{8\pi^2 I_e}$$

Critical Thinking Questions

1. Does O_2 have a permanent dipole moment? Will O_2 produce a rotational spectrum?

2. Will $CHCl_3$ produce a rotational spectrum?

3. Is the following rotational transition for a molecule such as CO allowed or forbidden: $J = 7 \rightarrow J = 6$?

Exercises

1. What are the two main components of dry air? Explain why air does not absorb microwave radiation in your microwave oven. Explain why water does absorb microwave radiation.

2. Which of the following molecules will not have a rotational spectrum: NH_3; BF_3; S_2; SO_2; CO_2; SO_3; $SiCl_4$?

3. Which of the following rotational transitions are allowed:
$J = 5 \rightarrow J = 4$; $J = 4 \rightarrow J = 5$; $J = 5 \rightarrow J = 0$; $J = 0 \rightarrow J = 5$?

Model 2: A Simulated Rotational Spectrum for a Diatomic Molecule.

The degeneracy of each energy level has been omitted for clarity in the diagram.

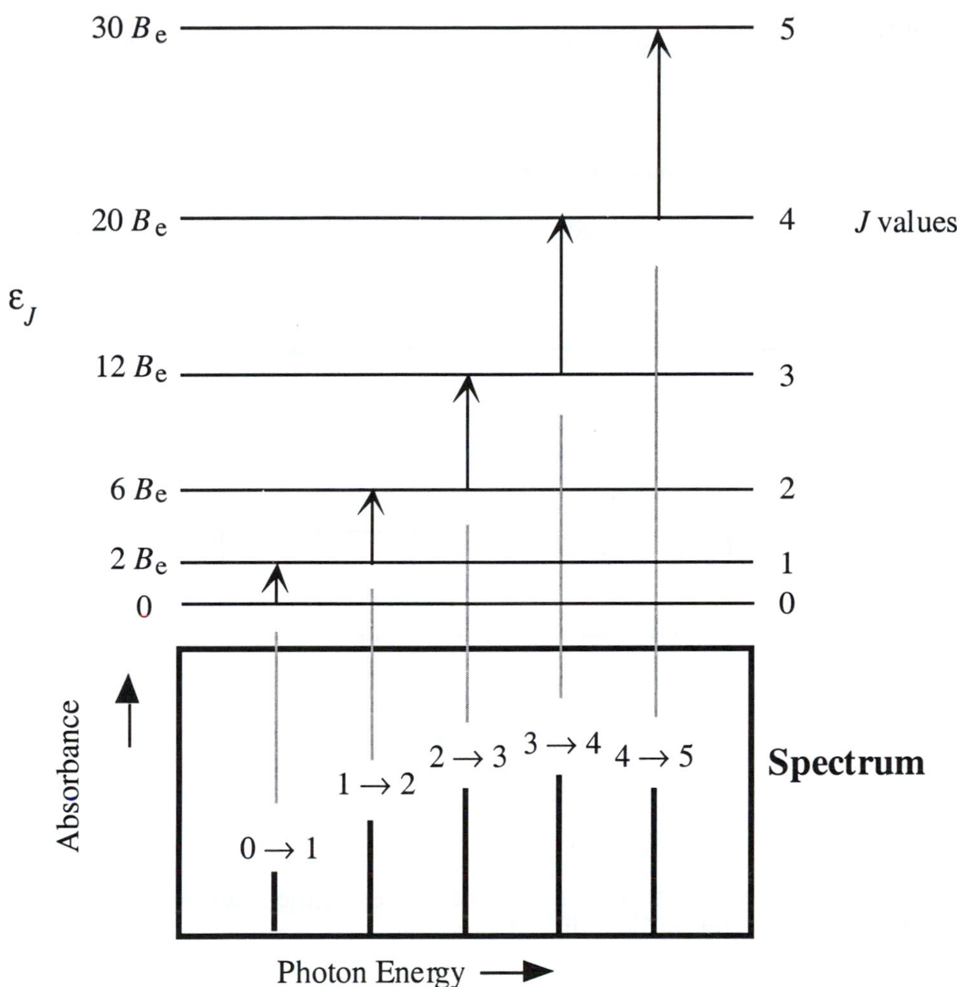

Critical Thinking Questions

4. Is energy absorbed or released for the rotational transition $J = 3 \rightarrow J = 4$?

5. A diatomic molecule undergoes a transition from rotational state J to rotational state $J + 1$. Derive a general expression that gives $\Delta\varepsilon$ for this rotational transition (in terms of B_e and J).

6. The transition $J = 0 \rightarrow J = 1$ occurs at $2B_e$. Label the photon energy of the other transitions.

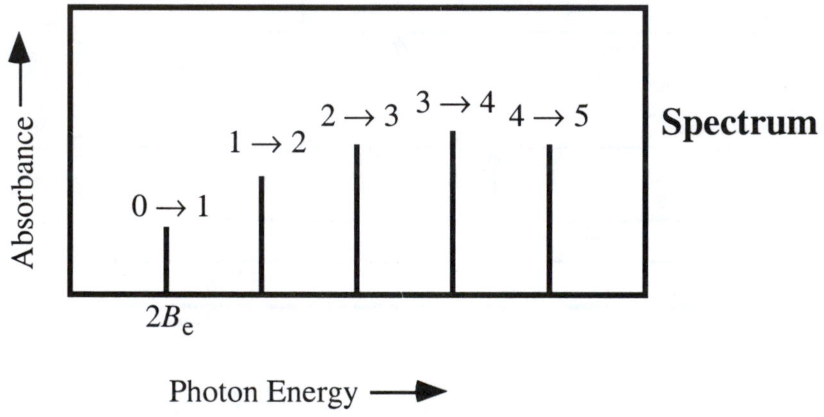

7. Describe how the spacing between adjacent lines of the microwave spectrum in CTQ 6 changes as the initial value of J increases.

8. Why is the $J = 3 \rightarrow J = 4$ transition more intense than the $J = 0 \rightarrow J = 1$ transition? [Hint : remember the Boltzmann distribution for rotational energies—see CTQ 17 of ChemActivity 20 on Selection Rules.]

Exercises

4. The wave number of the radiation absorbed in the $J = 0 \rightarrow J = 1$ transition of carbon monoxide has been measured as 3.842 cm^{-1}. Convert this wave number to joules. Calculate the wavelength of the photon. In what region of the electromagnetic spectrum is this photon? If a molecule in the $J = 1$ state loses energy and rotates with the energy given by $J = 0$, what is the energy, in joules, of the emitted photon? What is the moment of inertia of the CO molecule? If the absorption of this photon is from a molecule comprised of ^{12}C and ^{16}O, what is the bond length of the CO molecule?

5. Estimate, as precisely as you can, the energy, in cm^{-1}, for the radiation absorbed in the $J = 0 \rightarrow J = 1$ transition for $^{13}C^{16}O$. Explain your answer.

6. The high temperature microwave spectrum of KCl vapor shows an absorption at a frequency of 15,376 MHz. This frequency represents a photon with energy of 10.19×10^{-24} J. This peak has been identified with the $J = 1 \rightarrow J = 2$ transition of $^{39}K^{35}Cl$. Given that the atomic masses of ^{39}K and ^{35}Cl are 38.96 and 34.97 g/mole, respectively, calculate the internuclear distance of $^{39}K^{35}Cl$ in meters and pm.

7. The spacing between the lines in the microwave spectrum of $^{1}H^{35}Cl$ is 6.350×10^{5} M Hz. Calculate the internuclear distance of $^{1}H^{35}Cl$ in meters and pm. The atomic mass of ^{1}H is 1.008 g/mole. Compare this bond length to the bond length of $^{39}K^{35}Cl$. (See Exercise 6.) Does the comparison make sense?

Model 3: Selection Rules for Vibrational Transitions.

1. The vibrating molecule must produce an oscillating dipole moment (see CA 20). Homonuclear diatomic molecules do not increase their vibrational energy by absorption of IR radiation because they cannot have an oscillating dipole moment. All other molecules do have a vibrational spectrum (absorb in the IR).

2. For a harmonic oscillator-type molecule the selection rule is that

   $$\Delta\upsilon = \pm 1.$$

   The harmonic oscillator is only an approximation for the vibrational motion of a diatomic molecule. In addition to the *fundamental* transition, $\upsilon = 0 \rightarrow \upsilon = 1$, other transitions, called *overtones*, are sometimes observed, $\upsilon = 0 \rightarrow \upsilon = 2$ or $\upsilon = 0 \rightarrow \upsilon = 3$.

Recall for the harmonic oscillator model (diatomic molecules):

$$\varepsilon_\upsilon = (\upsilon + \frac{1}{2})\frac{h}{2\pi}\sqrt{\frac{k}{\mu}} \qquad \upsilon = 0, 1, 2, \ldots \qquad (2)$$

Critical Thinking Questions

9. Will O_2 produce a vibrational (IR) spectrum?

10. Recall the Boltzmann distribution for vibrational energy states and explain why transitions such as $\upsilon = 1$ to $\upsilon = 2$ are not normally observed at or near room temperature.

11. Explain why the symmetrical stretch of CO_2, shown below, cannot be excited by IR radiation.

$$\longleftarrow O{=}C{=}O \longrightarrow$$

Exercises

8. Explain why dry air does not absorb IR radiation. Explain why water does absorb IR radiation.

9. Will the bending mode of CO_2 , shown below, be excited by IR radiation?

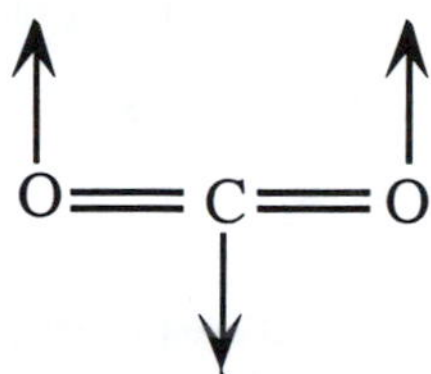

10. Will the antisymmetrical stretch of CO_2 , shown below, be excited by IR radiation?

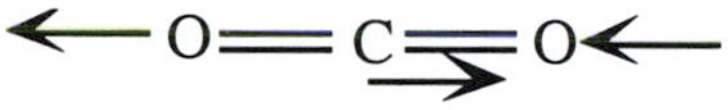

Model 4: A Simulated Rotational-Vibrational Spectrum for a Diatomic Molecule.

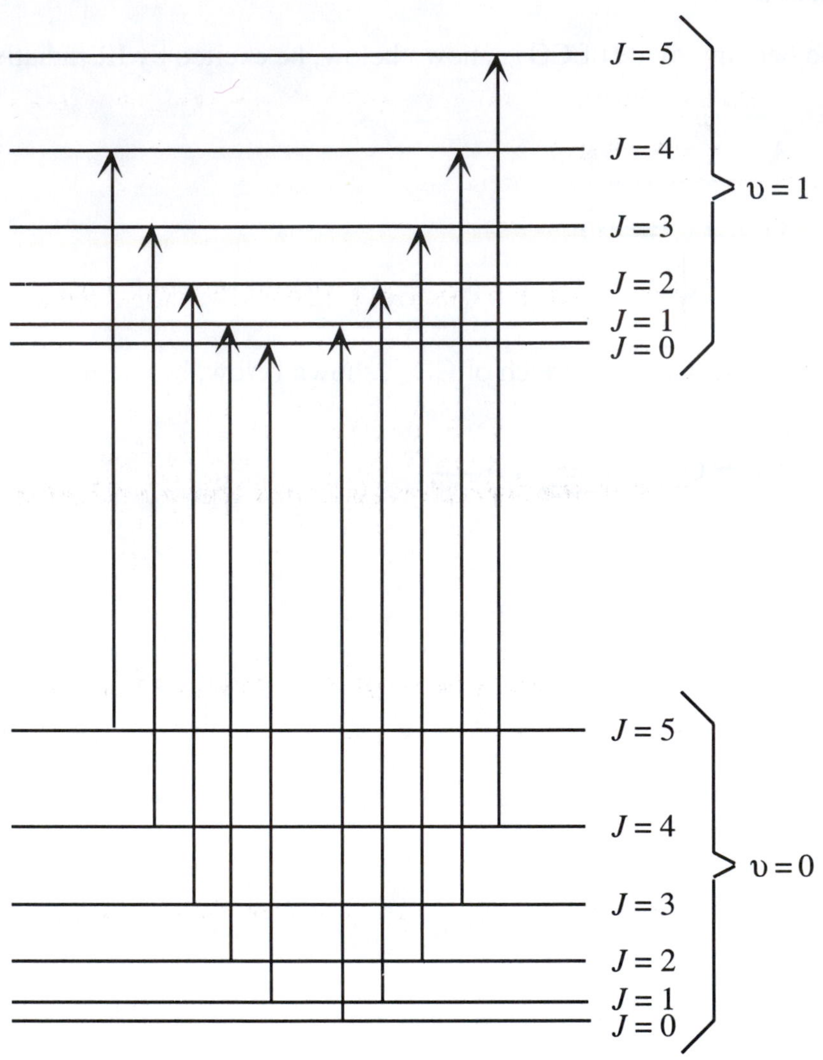

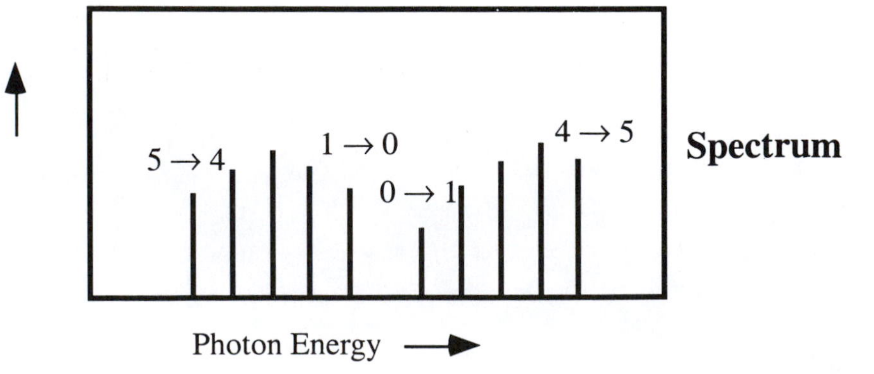

When a diatomic molecule absorbs sufficient energy to change the vibrational motion of the molecule it obviously has more than enough energy to change the rotational motion. It turns out that for an allowed vibrational transition $\Delta\upsilon = \pm 1$ *and* $\Delta J = \pm 1$. Thus, for a diatomic molecule, both the vibrational energy and the rotational energy change. The energy levels are given by:

$$\varepsilon_{\upsilon,J} = (\upsilon + \frac{1}{2})\,\frac{h}{2\pi}\sqrt{\frac{k}{\mu}} \;+\; B_e J(J+1) \tag{3}$$

$$\upsilon = 0, 1, 2, \ldots \text{ and } J = 0, 1, 2, \ldots \tag{4}$$

Figure 1. The fundamental, $\upsilon = 0$ to $\upsilon = 1$, absorption band of HCl. The lines are doublets because of the presence of two isotopes of chlorine, ^{35}Cl and ^{37}Cl. The transitions are indexed as follows: $J(\upsilon = 0) \rightarrow J(\upsilon = 1)$.

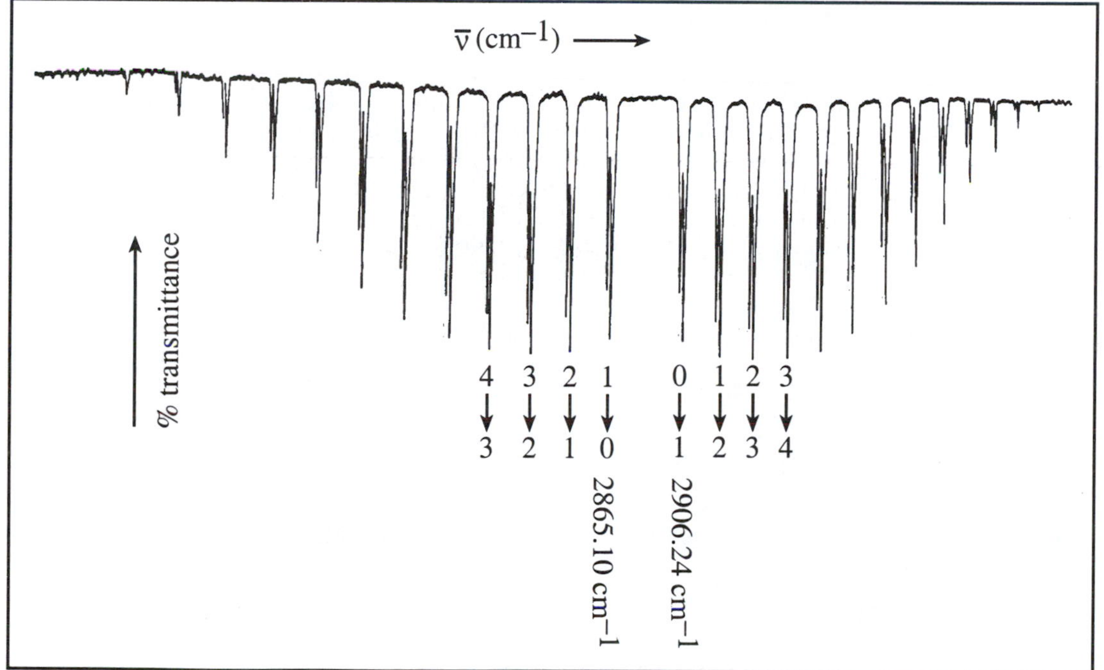

Critical Thinking Questions

12. The lines in the spectrum in Model 4 appear to be equally spaced, but there is a *gap* between the $\upsilon = 0$, $J = 0$ to $\upsilon = 1$, $J = 1$ transition and the $\upsilon = 0$, $J = 1$ to $\upsilon = 1$, $J = 0$ transition. Why?

Model 5: Observed vibrational transition energies of HCl and calculated vibrational transition energies based on the $\upsilon = 0$ to $\upsilon = 1$ transition and the harmonic oscillator model.

Transition	Description	$\Delta\varepsilon$ (cm^{-1}) Observed	Calculated[a]
$0 \rightarrow 1$	fundamental	2,885.9	
$0 \rightarrow 2$	first overtone	5,668.0	5,771.8
$0 \rightarrow 3$	second overtone	8,347.0	8,657.7
$0 \rightarrow 4$	third overtone	10,923.1	11,543.6
$0 \rightarrow 5$	fourth overtone	13,396.5	

[a]Based on the observed fundamental at 2,885.9 cm^{-1}, equation (2).

Critical Thinking Questions

13. Using the observed value for $\Delta\varepsilon$ of the fundamental transition, show that the calculated wave number (in Model 5) for the first overtone is correct.

14. Determine the calculated value for $\Delta\varepsilon$ for the fourth overtone.

15. According to Model 5, what is the *actual* energy of separation (in cm^{-1}) between:

    a) ε_2 and ε_1?

    b) ε_3 and ε_2?

    c) ε_4 and ε_3?

16. Are the vibrational energy levels equally spaced in HCl?

17. As the vibrational energy increases, do the energy level spacings increase or decrease?

18. Based on the data in Model 5 and your answers to CTQs 16 and 17, comment on the validity of the harmonic oscillator model in describing the vibrational potential energy of the HCl molecule.

Model 6: The Morse Potential.

In 1929, Morse proposed the following potential energy function for the vibrating diatomic molecule in an attempt to improve the correlation between spectroscopically observed and calculated vibrational energy levels.

$$V = D_e[1 - e^{-\beta(r - r_e)}]^2 \tag{5}$$

The Morse potential energy function is purely empirical and only three parameters are used:

- D_e, which is called the dissociation energy and is defined in Figure 2 (next page).
- r_e, the equilibrium internuclear distance.
- β, which is related to the force constant and the dissociation energy by the equation

$$\beta = \left(\frac{k}{2D_e}\right)^{1/2}$$

Solution of the Schrödinger equation yields wave functions (not given here) and the following energies:

$$\varepsilon_\upsilon = \omega_e\left(\upsilon + \frac{1}{2}\right) - \omega_e x_e\left(\upsilon + \frac{1}{2}\right)^2 \tag{6}$$

where ω_e is an experimental parameter which is almost equal to $\frac{h}{2\pi}\sqrt{\frac{k}{\mu}}$ [see the harmonic oscillator model] and $\omega_e x_e$ is an experimental parameter called the anharmonicity constant ($\omega_e x_e << \omega_e$). It can be shown that

$$D_e = \frac{\omega_e^2}{4\omega_e x_e} \tag{7}$$

Figure 2. The Morse potential and vibrational energy levels for the dihydrogen molecule, H_2.

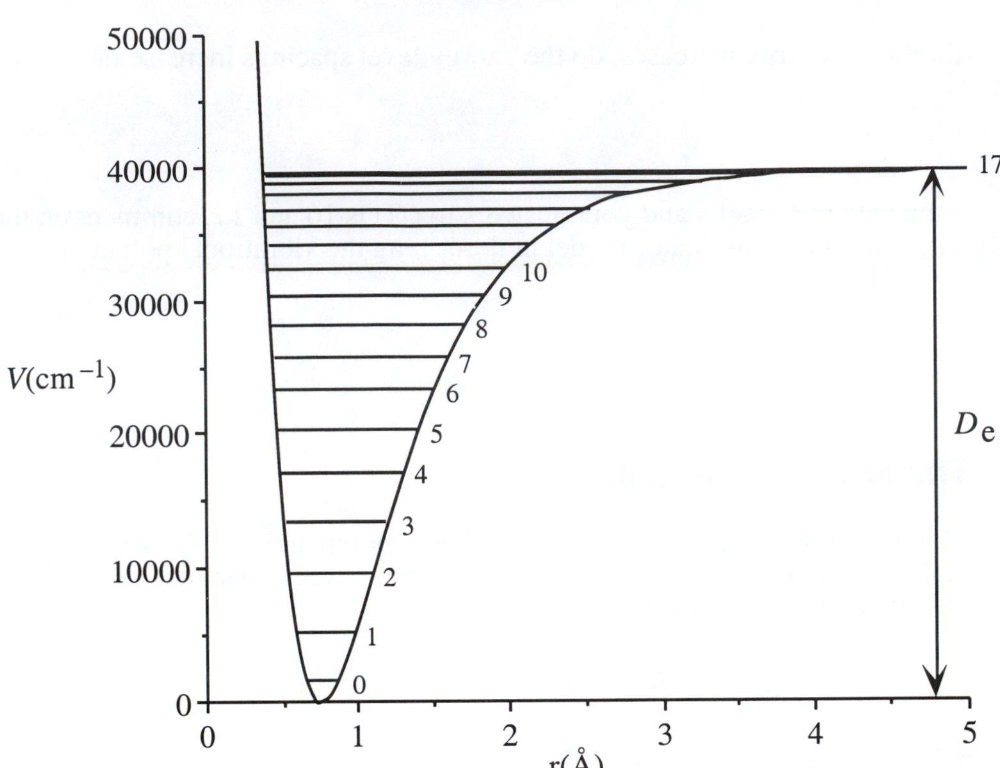

The agreement between the observed energy levels, D_e, r_e, and so on, is remarkable—but not exact. For examples: the Morse potential predicts 18 bound states ($\upsilon = 0$ to 17), but the H_2 molecule has only 15 ($\upsilon = 0$ to 14); the Morse potential D_e is 7.928×10^{-19} J, but the experimental D_e is 7.607×10^{-19} J.

Critical Thinking Questions

19. According to Figure 2, as the vibrational energy increases, do the energy level spacings increase or decrease?

20. According to Figure 2, what happens to an H_2 molecule in the $\upsilon = 18$ state?

Exercises

11. For H_2, $\omega_e = 4401.213$ cm^{-1} and $\omega_e x_e = 121.336$ cm^{-1}. Show that $D_e = 7.928 \times 10^{-19}$ J.

12. Calculate the wave number for the fundamental and the first overtone for H_2.

Model 7: The Vibrational Modes of Polyatomic Molecules.

Let the number of nuclei in a molecule equal N. One can specify the position of all of the nuclei with 3 N coordinates (x_i,y_i,z_i for each nucleus). The molecule is said to have $3N$ **degrees of freedom**. Alternatively, one could specify the center of mass of the entire *molecule* with three coordinates (X_i,Y_i,Z_i). Because the translational motion of a molecule is expressed through the motion of the center of mass of the molecule, a molecule is said to have three *translational degrees of freedom.* If the molecule is linear, the orientation of the nuclei about its center of mass requires two coordinates (usually θ and ϕ). A diatomic molecule is said to have two *rotational degrees of freedom.* If the molecule is not linear, the orientation of the nuclei about its center of mass requires three coordinates, and the molecule is said to have three rotational degrees of freedom. The remaining coordinates, $3N-5$ for a linear molecule and $3N-6$ for a nonlinear molecule, are assigned to the vibrational motion of the molecule and are called *vibrational degrees of freedom*.

	Linear	Nonlinear
Translational degrees of freedom	3	3
Rotational degrees of freedom	2	3
Vibrational degrees of freedom	$3N-5$	$3N-6$

Critical Thinking Questions

21. Why does glycine, H_2NCH_2COOH, have only three degrees of translational freedom?

22. Why does SO_2 have three degrees of rotational freedom whereas CO_2 has only two degrees of rotational freedom?

Exercise

13. Complete the following Table.

		Degrees of Freedom		
Molecule	Total	Translational	Rotational	Vibrational
N_2	6	3	2	1
HCl				
CS_2				
H_2O				
CH_4				
benzene				
H_2NCH_2COOH				

Information

It is beyond the scope of this course, but it can be shown (for the harmonic oscillator model) that the number of vibrational degrees of freedom of a molecule, N_{vib}, is expressed as N_{vib} harmonic oscillators or *vibrational modes* in the molecule. For example, the water molecule has $N_{vib} = 3$ and water has three vibrational modes; each of the modes has an oscillating dipole moment, and all three are *infrared active*.

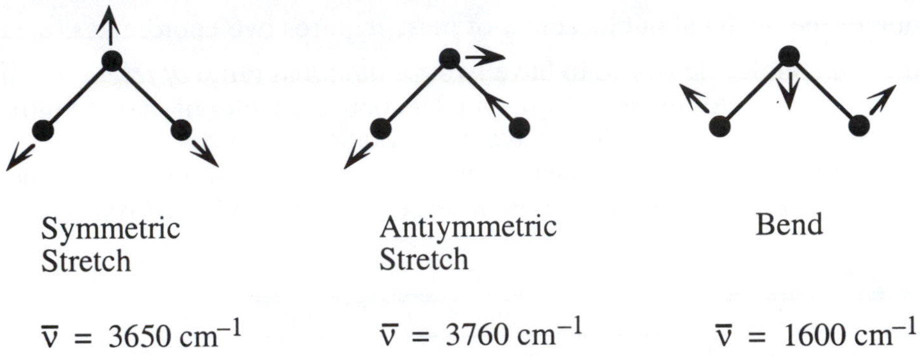

Problem

1. CO_2 has the four vibrational modes shown below.

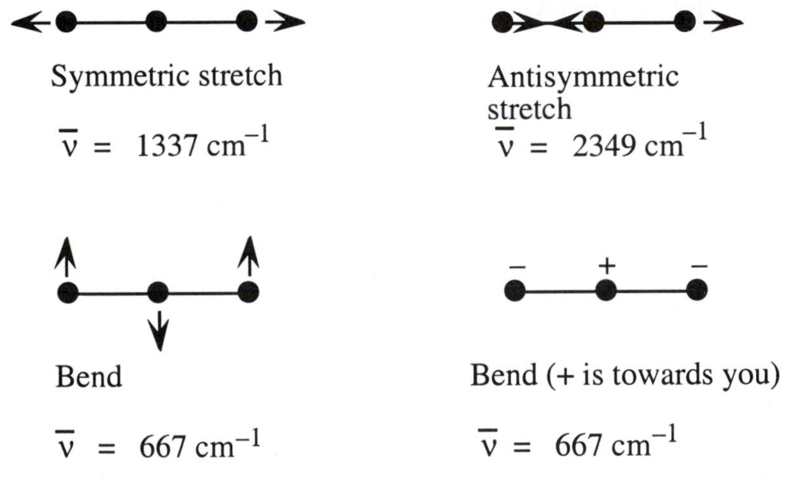

(a) Which modes are infrared active?

(b) The symmetric stretch, the antisymmetric stretch, and the bending mode on the left in the figure are all within the plane of the paper. The bending mode on the right in the figure is necessary because the molecule does not, of course, simply vibrate in a plane. However, why do the two bending modes have the identical wavenumbers?

(c) Sketch the IR spectrum for CO_2.

ChemActivity **22**

Electronic Spectra of Atoms and Molecules

(Where did that flash of light come from?)

We have seen that molecules can change rotational energy and vibrational energy by absorption of electromagnetic radiation in the microwave and infrared regions of the spectrum. Now, we turn our attention to changes in the electronic energy of atoms and molecules. In the activity we examine questions such as:

- How does a molecule in a particular electronic state, say $^1\Sigma_g$, undergo a transition that takes the molecule to a higher energy electronic state?
- What happens to a molecule in an electronically excited state?

Model 1: Selection Rules for Electronic Transitions in Atoms.

1. $\Delta S = 0$

2. $\Delta L = \pm 1$

Critical Thinking Questions

1. Is the following transition allowed or forbidden: ${}^2D \rightarrow {}^2P$? Explain.

2. Is the following transition allowed or forbidden: ${}^3D \rightarrow {}^1P$? Explain.

Exercises

1. What is the term symbol for the $1s^1$ configuration of the hydrogen atom? What is the term symbol for the $2s^1$ configuration of the hydrogen atom? What is the term symbol for the $2p^1$ configuration of the hydrogen atom? Is the transition from $1s^1$ to $2s^1$ allowed or forbidden? Is the transition from $2s^1$ to $1s^1$ allowed or forbidden? Is the transition from $2p^1$ to $1s^1$ allowed or forbidden?

2. Sodium emits a yellow line in a flame that arises from an electronic transition from $3p^1$ to $3s^1$. Is this an allowed or forbidden transition?

Model 2: Selection Rules for Electronic Transitions in Molecules.

1. The spin quantum number must not change, $\Delta S = 0$.
2. Other selection rules depend on the symmetry of the molecule. For example, for molecules with a center of symmetry, electronic transitions between ungerade and gerade states are allowed ($u \leftrightarrow g$) ; transitions between gerade states ($g \leftrightarrow g$) and transitions between ungerade states ($u \leftrightarrow u$) are forbidden.
3. There is no restriction on the change in the vibrational quantum number. In general, there is no restriction on changes in orbital angular momentum.

Critical Thinking Question

3. The ground state of O_2 is $^3\Sigma_g$. To which of the following excited states of O_2 is the ground state to excited state transition allowed?

$^1\Delta_g$ $^1\Sigma_g$ $^3\Sigma_u$ $^3\Sigma_g$

Exercises

3. Evaluate $\int_{-\infty}^{\infty} f(x)dx$ for each of the following $f(x)$; it is only necessary to determine if the integral is *zero* or *not zero*: x; x^2; x^3; x^4; x^5; x^6. Based on your result, predict the value of $\int_{-\infty}^{\infty} x^{27}$ without explicitly evaluating the integral. State a general rule for these types of integral.

4. Recall that the probability of a transition is proportional to the square of the transition dipole moment:

$$\text{transition dipole moment} = \int \Psi_a^* \hat{u} \Psi_b d\tau$$

The transition dipole moment for a transition from a gerade (even) state to a gerade (even) state is

$$\text{transition dipole moment} = \int \Psi_g^* \hat{u} \Psi_g d\tau$$

The dipole moment for a molecule is inherently ungerade (odd) because inversion of the dipole moment vector through the center of the molecule will always yield a vector that is the negative of the dipole moment. Recognize that the dipole moment operator is also inherently ungerade (odd), and use your rule developed in Exercise 3 to show that the transition dipole moment for this transition is zero and, therefore, that transitions of this type are forbidden. Similarly, show that the transition dipole moment for a transition from a gerade state to an ungerade state is not zero and, therefore, transitions of this type are allowed.

5. Is an atomic *s* orbital ungerade or gerade? Is an atomic *p* orbital ungerade or gerade? Show that the transition from a 2*s* orbital to a 1*s* orbital is forbidden. Show that the transition from a 2*p* orbital to a 1*s* orbital is allowed.

6. Which of the following transitions are allowed?
   (a) $3d^1 \rightarrow 4d^1$
   (b) $3d^1 \rightarrow 4p^1$
   (c) $3d^1 \rightarrow 4s^1$
   (d) ${}^1\Delta_g \rightarrow {}^1\Delta_g$
   (e) ${}^1\Delta_g \rightarrow {}^1\Pi_g$
   (f) ${}^3\Delta_g \rightarrow {}^3\Delta_u$
   (g) ${}^3\Delta_g \rightarrow {}^1\Delta_u$

Model 3. The Franck-Condon Principle.

An electron is much lighter than any nucleus and, as a result, an electron (in an atom or molecule) moves much faster than any nucleus. Thus, during the time when an electronic transition occurs, there is no appreciable change in the internuclear distances in the molecule—this is called the **Franck-Condon Principle**.

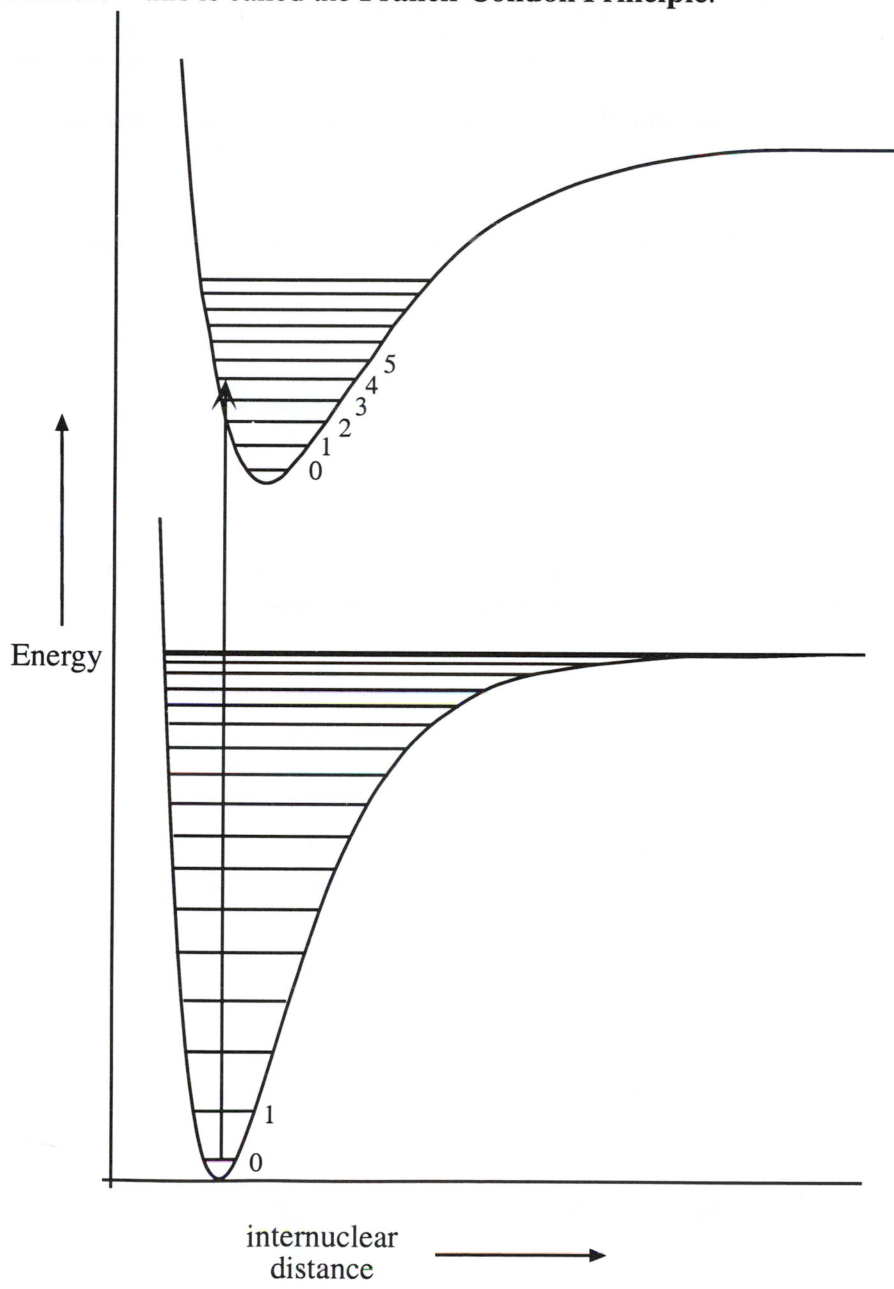

Critical Thinking Questions

4. Identify that part of the figure in Model 3 which represents (a) the electronic ground state and (b) the electronic excited state.

5. According to the figure in Model 3, which has the greater equilibrium internuclear distance in the lowest vibrational state–the ground state molecule or the excited state molecule?

6. What does your answer to CTQ 5 suggest about the relative bond strengths in the ground and excited states?

7. Based on the relative depths of the potential wells, predict which state has the weaker bond in Model 3—the excited state or the ground state.

8. Recall the Boltzmann distribution for vibrational energy levels and explain why at room temperature the most likely transition is from the $\upsilon = 0$ state of the electronic ground state (rather than the $\upsilon = 1$ state of the electronic ground state).

9. Explain why the most likely transition begins at r_e in the $\upsilon = 0$ level of the electronic ground state.

10. Draw a straight line from r_e ($\upsilon = 0$) of the electronic ground state to r_e ($\upsilon' = 0$) of the electronic excited state. Is this transition possible? Explain.

Information

Prior to an electronic transition, most molecules are in the $\upsilon = 0$ vibrational state for which the most probable internuclear distance is r_e. Because the Franck-Condon principle requires that any transition occurs without a change in internuclear distance, the maximum absorption intensity will correspond to a transition to the excited vibrational level with a most probable internuclear distance also at r_e (directly above r_e of the ground state). More precisely, the transition probability is given by the overlap integral of the initial and final vibrational wave functions.

Exercises

7. A UV absorption band near 140 nm in the low temperature spectrum of CO consists of a series of lines. The five lowest energy lines are given below. It has been established that the 64,703 cm^{-1} line arises from a transition to the $\upsilon' = 0$ vibrational level in the excited state. (a) Draw a diagram, similar to the one in Model 3, that illustrates these transitions. (b) Fill in the υ (ground state) and υ' (excited state) quantum numbers in the table below. (c) Use the harmonic oscillator model and calculate the force constant for the excited state bond from the energy spacing between $\upsilon' = 0$ and $\upsilon' = 1$.

Transition Wavenumber (cm^{-1})	υ	υ'
64,703	0	0
66,231		
67,675		
69,088		
70,470		

8. The force constant for CO in textbooks is typically reported as 1854 N m^{-1}. Compare this value with the value obtained in Exercise 7. Account for the rather large discrepancy.

Model 4: Fluorescence.

When an electronically excited state molecule releases a photon in a radiative transition to the ground electronic state and the spin quantum number, S, does not change, the molecule is said to fluoresce and the process is called fluorescence. Typically, fluorescence occurs about 10^{-6} seconds after absorption of a photon.

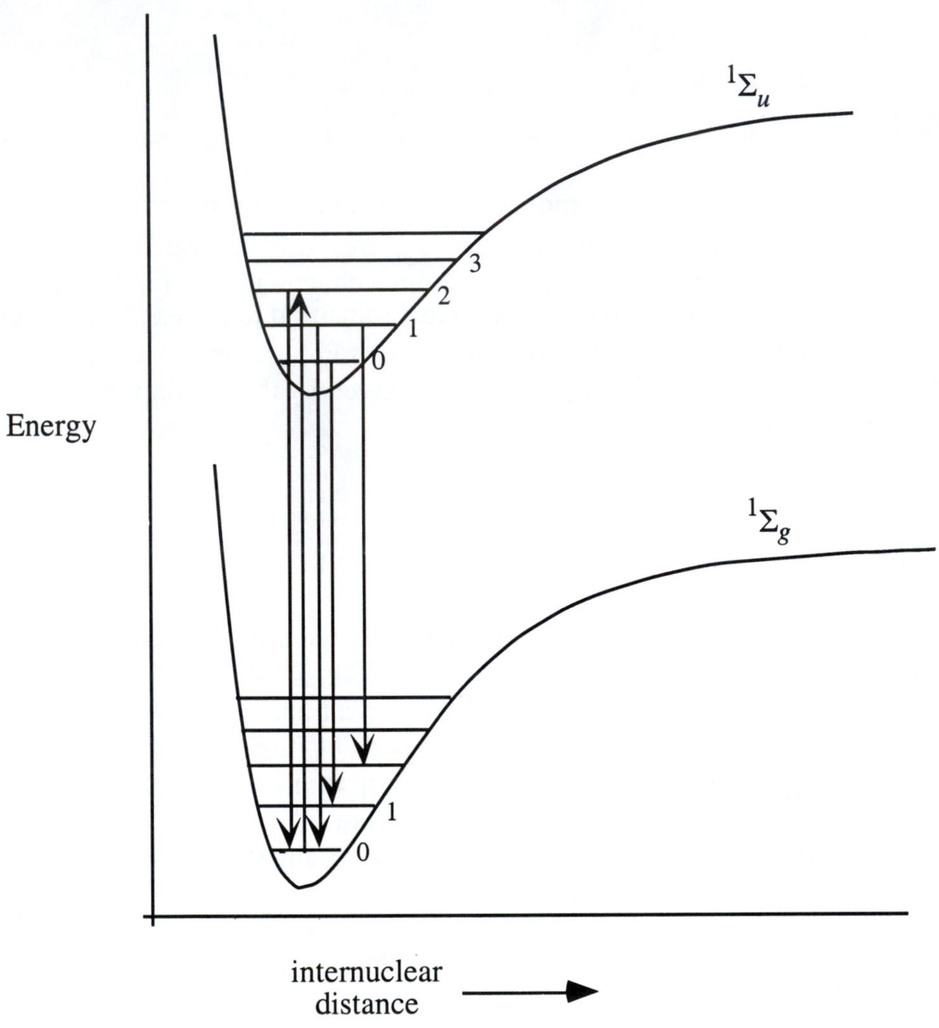

Critical Thinking Questions

12. Which transition(s) in Model 4 are absorptions?

13. Which transition(s) in Model 4 are emissions?

14. What is the value of the quantum number *S* for excited state in in Model 4?

15. What is the value of the quantum number *S* for ground state in in Model 4?

16. What is ΔS for the fluorescence in in Model 4?

17. Do the emissions in in Model 4 (fluorescence) represent allowed or forbidden transitions?

Information

When a ground state molecule absorbs a photon (say from $\upsilon = 0$ in the ground state to $\upsilon' = 5$ in the excited electronic state) several subsequent processes can result in fluorescence:

- quickly emit a UV/VIS energy photon and return to the ground state.
- emit one or more infrared photon(s) (and go to $\upsilon' \leq 4$ of the excited state) and then emit a UV/VIS energy photon and return to the ground state.
- undergo a radiationless transition to a lower vibrational state (to $\upsilon' \leq 4$ of the excited state). The vibrational energy lost can be transferred (a) to another molecule in a collision, (b) to a different vibrational mode within the same molecule, (c) to rotational motion within the same molecule, or (d) any combination of the above. Once the molecule is in a lower vibrational energy state ($\upsilon' \leq 4$ of the excited state), the molecule can emit a UV/VIS energy photon and return to the ground state. This is the predominant fluorescence process for molecules in liquid and solid phases.

Critical Thinking Question

18. Based on Model 4 and the information above, explain why the energy of the emitted photon (fluorescence) is generally lower than the energy of the absorbed photon.

Information

A functional group or other group of atoms within a molecule which exhibits a characteristic absorption in the UV/VIS is called a *chromophore*. Most organic compounds with a carbonyl group absorb near 200 nm. This absorption is attributed to a π to π^* transition. Another absorption (in carbonyl compounds) is found near 300 nm and is attributed to a nonbonding (lone pair) to π^* transition.

Critical Thinking Question

19. If a carbonyl chromophore absorbs at 200 nm, is the wavelength of the fluorescence more likely to be 220 nm or 180 nm?

Problem

1. The UV absorption spectrum of pyrazine vapor at 20°C produces a set of almost equally spaced lines. When the spectrum is taken at 60°C the same lines are seen but additional lines appear at *lower* energies. Explain the origin of the additional lines in the spectrum at 60°C.

Model 5: Phosphorescence.

Another possibility exists for an electronically excited molecule. The excited molecule can undergo a collision (radiationless transition) that causes the spin quantum number to change (usually, this is from an $S = 0$ state to an $S = 1$ state); this is called **intersystem crossing**. The transition back to the ground state is now forbidden and the molecule is *trapped* in the excited state. It will, of course, lose vibrational energy and most likely wind up in the $\upsilon = 0$ vibrational level of the excited state. Eventually, the electronically excited state molecule releases a photon in a radiative transition to the ground electronic state and the spin quantum number, S, *does* change, the molecule is said to **phosphoresce** and the process is called **phosphorescence**. Phosphorescence typically occurs at 10^{-3} to 10 seconds after absorption of a photon.

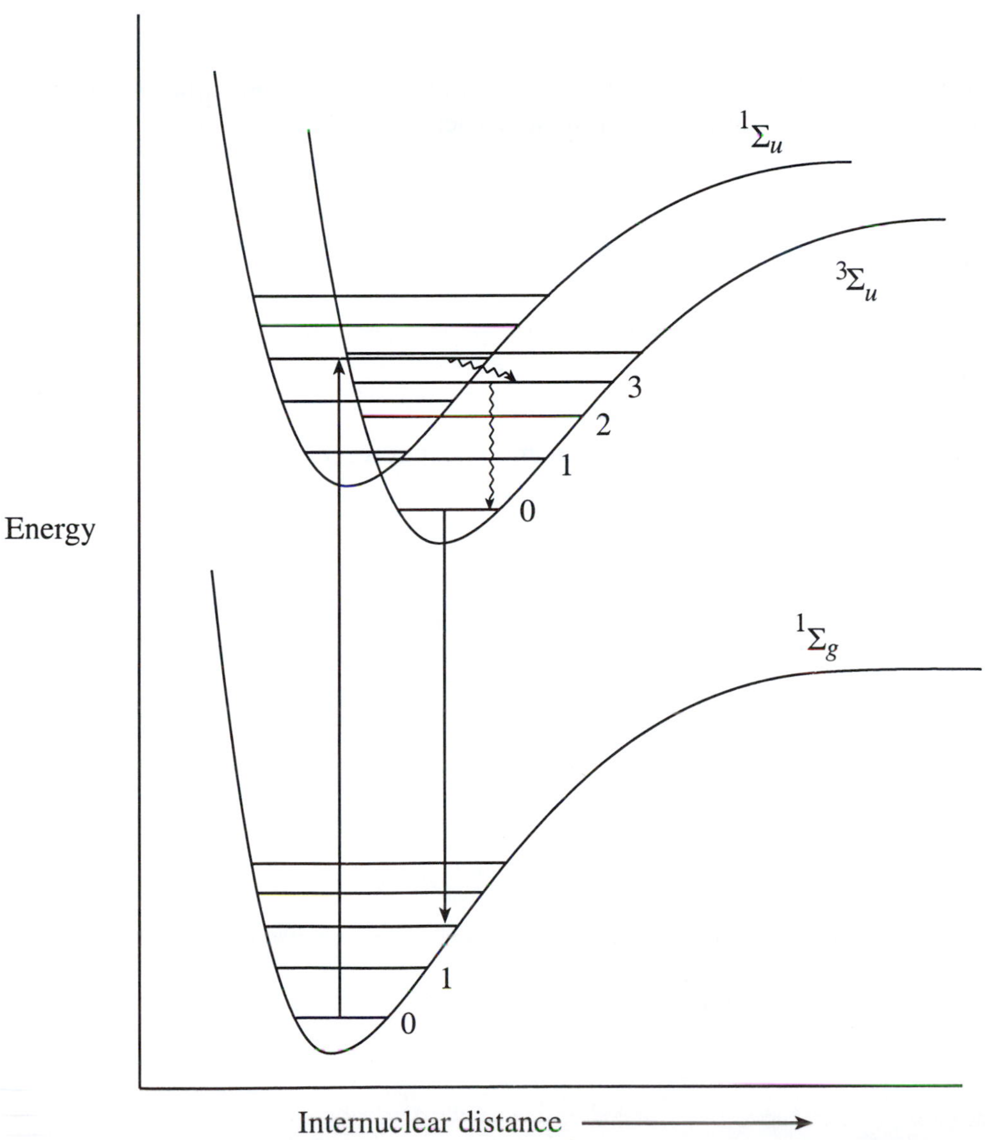

Critical Thinking Questions

20. For each of the following processes, indicate which arrow in Model 5 corresponds to which process: absorption, phosphorescence, intersystem crossing, vibrational relaxation.

21. In Model 5, at which value of υ' in the singlet state does the intersystem crossing begin? At which value of υ' in the triplet state does the intersystem crossing end?

22. According to Model 5, is the energy of the emitted photon (phosphorescence) higher or lower than the energy of the absorbed photon?

Problem

2. Describe, in some detail, fluorescence and phosphorescence. Include a diagram to illustrate each.

ChemActivity 23

Photoelectron Spectra of Molecules

(Is this bad news for valence bond theory?)

Valence bond theory (localized **two-electron bonds** plus, in some cases, **resonance structures**) is extremely useful in chemistry. It is a rather simple task to construct Lewis structures for molecules (no computer is required), and these structures are easily used to predict molecular shapes and bond angles. Often, Lewis structures enable predictions of bond length and bond strength (a C–C double bond is stronger and shorter than a C–C single bond). Resonance structures can be used to predict the relative stability of molecules and various intermediates. A great deal of information can be gained at relatively little expense.

Molecular orbital theory (delocalized bonding throughout the entire molecule) requires extensive calculations; the up-front costs are higher in MO theory than valence bond theory. But the payoff is greater as well. Valence bond theory can accurately predict that the bond angle of H_2O is somewhat less than 109.45°, but it does not give a precise value. Valence bond theory can predict that napthalene has more resonance stabilization energy than benzene, but it does not predict by how many kJ/mole. MO theory permits calculation of the total electronic energy of a molecule, the energy levels of various orbitals, partial charges, bond lengths and angles, the energies of various excited states, and so on.

In this activity we make a comparison between the experimental photoelectron spectra of molecules and those predicted by valence bond theory and molecular orbital theory.

Model 1: The MOPAC/AM1 Description of CH_4.

Table 1. MOPAC Energy Levels for CH_4.

Energy Level	Energy (eV)	Energy Level	Energy (eV)
1	-28.88	5	4.66
2	-13.31	6	4.66
3	-13.31	7	4.66
4	-13.31	8	5.13

Table 2. MOPAC Wave functions for CH_4.

Atomic Orbital	Coefficients for each MO (Ψ_i) Ψ_1	Ψ_2	Ψ_3	Ψ_4
2*s*C1	-0.7843	0.00000	0.0000	-0.0002
2*px*C1	0.0000	0.0552	-0.7048	0.0787
2*py*C1	0.0000	-0.0367	0.0760	0.7063
2*pz*C1	0.0000	0.7082	0.0588	0.0305
1*s*H2	-0.3102	-0.0313	0.0651	0.6044
1*s*H3	-0.3102	0.0549	-0.5904	-0.1378
1*s*H4	-0.3102	0.4831	0.3036	-0.2120
1*s*H5	-0.3102	-0.5067	0.2215	-0.2544

Atomic Orbital	Coefficients for each MO (Ψ_i) Ψ_5	Ψ_6	Ψ_7	Ψ_8
2*s*C1	0.0008	-0.0001	0.0000	0.6204
2*px*C1	-0.0094	0.7012	0.0468	0.0002
2*py*C1	-0.7023	-0.0076	-0.0264	0.0010
2*pz*C1	-0.0258	-0.0471	0.7008	0.0000
1*s*H2	0.6151	0.0068	0.0230	-0.3928
1*s*H3	-0.1979	-0.5817	-0.0463	-0.3919
1*s*H4	-0.1912	0.3212	-0.4899	-0.3919
1*s*H5	-0.2279	0.2539	0.5131	-0.3919

Figure 1. The MO energy level diagram for methane.

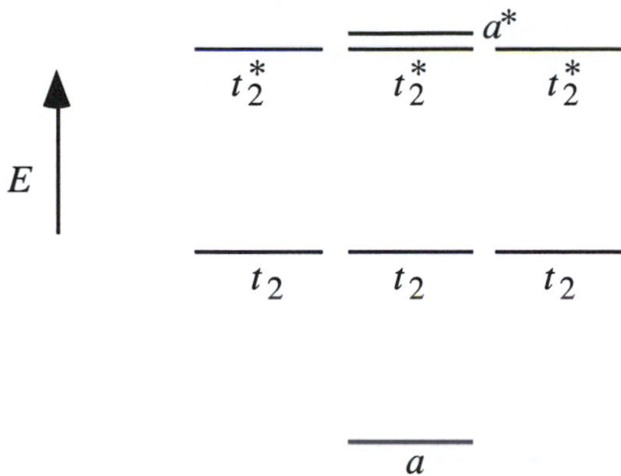

The labels "a" and "t_2" refer to various symmetry properties of orbitals.

Critical Thinking Questions

1. According to Table 1, what is the ionization energy of the MO labeled a in Figure 1?

2. According to Table 1, what is the ionization energy of the MOs labeled t_2 in Figure 1?

3. Place the appropriate number of electrons for methane in the diagram of Figure 1. Which statement is true?

   All of the electrons are bonding.
   Some of the electrons are bonding and some are antibonding.
   All of the electrons are antibonding.

4. Make a sketch of the photoelectron spectrum of methane based the MO description of methane (ignore the carbon $1s$ electrons).

Model 2: The Valence Bond (VB) Description of CH_4.

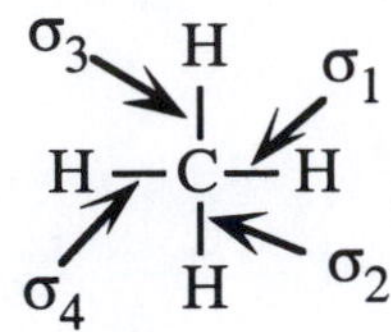

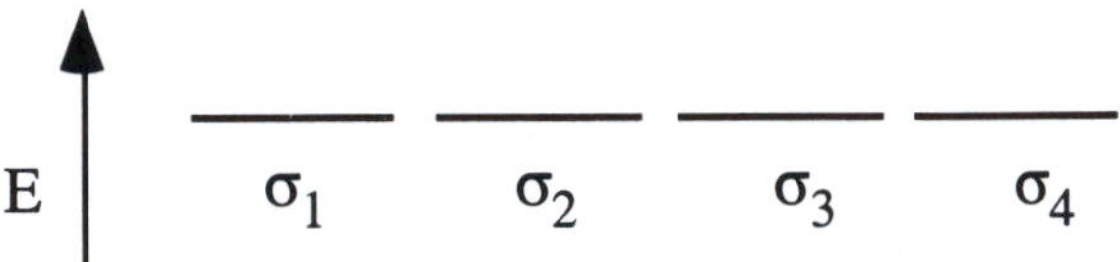

Each of the four σ energy levels (and valence bonds) is the result of the overlap of a carbon atom sp^3 hybrid orbital and a hydrogen atom *s* atomic orbital.

Critical Thinking Questions

5. Place the appropriate number of electrons for methane in Model 2. Which statement is true?

   All eight of the electrons are bonding.
   Some of the electrons are bonding and some are antibonding.
   All of the eight electrons are antibonding.

6. According to the valence bond (VB) description of methane, are all eight electrons at the same energy?

7. Make a sketch of the photoelectron spectrum of methane based the VB description of methane (ignore the carbon 1*s* electrons).

Model 3: The Photoelectron Spectrum of CH_4.

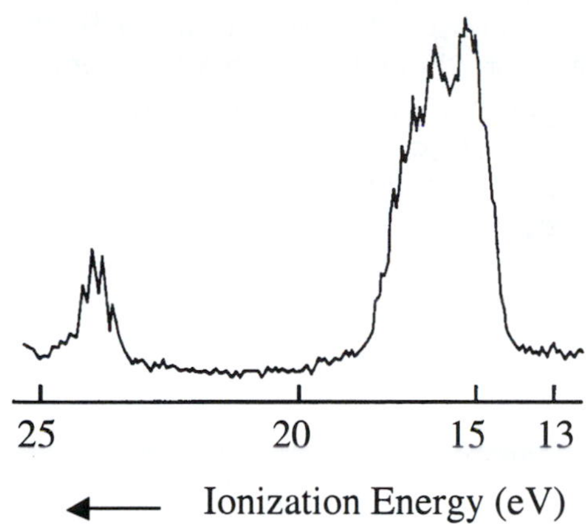

From A.W. Potts and W.C. Price, *Proc. Roy. Soc. (London) A*, *326*, 165(1972).

The lines are broad and a bit jagged because the ionization occurs from many different vibrational levels of the molecules. The peak corresponding to the ionization of the carbon 1*s* electrons, which occurs at 296 eV, is not shown.

Critical Thinking Questions

8. Based on the PES of methane, which is a better description of methane, the MO description or the VB description? Explain your answer.

Exercises

1. (a) The MOPAC/AM1 energy levels (in eV) for ammonia are: -32.68; -15.90; -15.90; -10.42; 4.22; 6.16; 6.16. Sketch the MO energy level diagram and place the correct number of electrons into the diagram. Sketch the PES of ammonia based on the MO model (ignore the nitrogen 1*s* electrons). (b) Write the Lewis structure for ammonia and use VB theory to describe the ammonia molecule. Make an energy level diagram for ammonia based on VB theory (similar to the diagram of Model 2). Sketch the PES of ammonia based on this model (ignore the nitrogen 1*s* electrons). What feature of the PES is different in the two models? (c) The PES of ammonia has three peaks (ignoring the nitrogen 1*s* electrons) in a ratio of 2:4:2 (high *IE* to low *IE*), which is the better theory for predicting the energy levels in ammonia—MO theory or VB theory?

2. (a) Examine the MO diagram of ethene (see ChemActivity 16; MOPAC). Sketch the PES of ethene based on this model (ignore the carbon 1*s* electrons). (b) Write the Lewis structure for ethene and use VB theory to describe the ethene molecule. Make an energy level diagram for ethene based on VB theory. Sketch the PES of ethene based on this model (ignore the carbon 1*s* electrons). What feature of the PES is different in the two models? (c) The PES of ethene is shown below. Based on the actual spectrum, which is the better theory for predicting the energy levels in ethene—MO theory or VB theory?

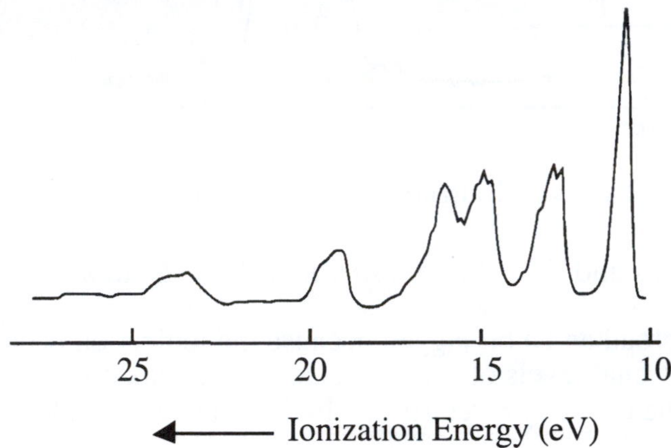

From C.R. Brundle, M.B. Robin, H. Basch, M. Pinsky, and A. Bond, *J. Amer. Chem. Soc.*, *92*, 3863(1970).

Appendix

TABLE A.1 Values of Selected Fundamental Constants

Speed of light in a vacuum (c)	$c = 2.99792458 \times 10^8$ m s^{-1}
Charge on an electron (q_e)	$q_e = 1.6021892 \times 10^{-19}$ C
Rest mass of an electron (m_e)	$m_e = 9.109534 \times 10^{-28}$ g
	$m_e = 5.4858026 \times 10^{-4}$ amu
Rest mass of a proton (m_p)	$m_p = 1.6726485 \times 10^{-24}$ g
	$m_p = 1.00727647$ amu
Rest mass of a neutron (m_n)	$m_n = 1.6749543 \times 10^{-24}$ g
	$m_n = 1.008665012$ amu
Planck's constant (h)	$h = 6.626176 \times 10^{-34}$ J s
Planck's constant ($\hbar$)	$\hbar = 1.054589 \times 10^{-34}$ J s
Ideal gas constant (R)	$R = 0.0820568$ L atm mol^{-1} K^{-1}
	$R = 8.31441$ J mol^{-1} K^{-1}
Atomic mass unit (amu)	1 amu = $1.6605655 \times 10^{-24}$ g
Boltzmann's constant (k)	$k = 1.380662 \times 10^{-23}$ J K^{-1}
Avogadro's constant (N)	$N = 6.022045 \times 10^{23}$ mol^{-1}
Rydberg constant (R_H)	$R_H = 1.09737318 \times 10^7$ m^{-1}

TABLE A.2 Selected Conversion Factors

Energy	1 J = 0.2390 cal = 10^7 erg 1 cal = 4.184 J (by definition) 1 ev/atom = $1.6021892 \times 10^{-19}$ J atom^{-1} = 96.484 kJ mol^{-1}
Temperature	K = °C + 273.15
Pressure	1 atm = 760 mm Hg = 760 Torr = 101.325 kPa
Mass	1 kg = 2.2046 lb 1 lb = 453.59 g = 0.45359 kg
Volume	1 mL = 0.001 L = 1 cm^3 (by definition) 1 qt = 0.946326 L 1 L = 1.05672 qt
Length	1 mi = 1.60934 km 1 in = 2.54 cm (by definition)

TABLE A.3 The Hermite Polynomials, $H_v(x)$.

$$H_0(x) = 1$$

$$H_1(x) = 2x \qquad x = z\sqrt{a}$$

$$H_2(x) = \; = 4x^2 - 2$$

All Hermite polynominals can be generated by the formula:

$$H_v(x) = (-1)^v e^{-x^2} \frac{dn}{dx^v}(e^{x^2})$$

TABLE A.4 The Associated Legendre Polynomials.

(The associated Legendre polynomials are derivatives of the Legendre polynomials.)

$$\Theta_{Jm} = f(\theta, J, m) \qquad J = 0, 1, 2, \ldots \qquad m = 0, \pm1, \pm2, \pm3, \ldots, \pm J$$

$$\Theta_{00} = \frac{\sqrt{2}}{2} \qquad \Theta_{10} = \frac{\sqrt{6}}{2}\cos(\theta)$$

$$\Theta_{11} = \Theta_{1,-1} = \frac{\sqrt{3}}{2}\sin(\theta)$$

$$\Theta_{20} = \frac{\sqrt{10}}{4}(3\cos^2(\theta) - 1)$$

$$\Theta_{21} = \Theta_{2,-1} = \frac{\sqrt{15}}{2}\sin(\theta)\cos(\theta)$$

$$\Theta_{22} = \Theta_{2,-2} = \frac{\sqrt{15}}{4}\sin^2(\theta)$$

TABLE A.5 Some Useful Integrals

$$\int (\sin mx)(\sin nx)\, dx = \frac{\sin(m-n)x}{2(m-n)} - \frac{\sin(m+n)x}{2(m+n)} \qquad \text{for } m^2 \neq n^2$$

$$\int (\sin^2 ux\, dx) = \frac{x}{2} - \frac{1}{4u}\sin 2ux$$

$$\int x(\sin^2 ux\, dx) = \frac{x^2}{4} - \frac{x \sin 2ux}{4u} - \frac{\cos 2ux}{8u^2}$$

TABLE A.6 The General Formula for Hydrogen Atom Wavefunctions.

$$R_{n,\ell} = -\sqrt{\frac{4(n-\ell-1)!}{n^4 a_o^3\,[(n+\ell)!]^3}}\left(\frac{2r}{na_o}\right)^{\ell} e^{-r/na_o}\,\mathrm{L}(x)$$

$n = 1, 2, 3, \ldots$

$\ell = 0, 1, 2, \ldots, n-1$

where $x = \dfrac{2r}{na_o}$

and $\mathrm{L}(x) = \dfrac{\partial^{2\ell+1}}{\partial x^{2\ell+1}}\left(e^{x}\dfrac{\partial^{n}}{\partial x^{n}}\,(x^{n}e^{-x})\right)$

$$\Theta_{\ell,m} = \sqrt{\frac{(2\ell+1)\,(\ell-m)!}{2(\ell+m)!}}\;\sin^{m}\theta\;\frac{1}{2^{\ell}\ell!}\;\frac{\partial^{m+\ell}}{\partial\mu^{m+\ell}}\,(\mu^2-1)^{\ell}$$

$m = 0, \pm1, \pm2, \ldots, \pm\ell$

$\mu = \cos\theta$

$\Phi_m = \dfrac{1}{\sqrt{2\pi}}\,e^{im\phi} \qquad m = 0, \pm1, \pm2, \ldots, \pm\ell$

$$\Psi_{n,\ell,m} = R_{n,\ell}\,\Theta_{\ell,m}\,\Phi_m$$

TABLE A.8 Wavefunctions for Hydrogen-like Atoms.

$$\Psi_{1s} = \Psi_{100} = \frac{1}{\pi^{1/2}}\left(\frac{Z}{a_o}\right)^{3/2} e^{-Zr/a_o}$$

$$\Psi_{2s} = \Psi_{200} = \frac{1}{4(2\pi)^{1/2}}\left(\frac{Z}{a_o}\right)^{3/2}\left(2 - \frac{Zr}{a_o}\right) e^{-Zr/2a_o}$$

$$\Psi_{210} = \Psi_{2p_z} = \frac{1}{4(2\pi)^{1/2}}\left(\frac{Z}{a_o}\right)^{5/2} r\, e^{-Zr/2a_o} \cos\theta$$

$$\Psi_{2p_x} = \frac{1}{4(2\pi)^{1/2}}\left(\frac{Z}{a_o}\right)^{5/2} r\, e^{-Zr/2a_o} \sin\theta\cos\phi$$

$$\Psi_{2p_y} = \frac{1}{4(2\pi)^{1/2}}\left(\frac{Z}{a_o}\right)^{5/2} r\, e^{-Zr/2a_o} \sin\theta\sin\phi$$

$$\Psi_{3s} = \Psi_{300} = \frac{1}{81(3\pi)^{1/2}}\left(\frac{Z}{a_o}\right)^{3/2}\left(27 - \frac{18Zr}{a_o} + \frac{2Z^2r^2}{a_o^2}\right) e^{-Zr/3a_o}$$

$$\Psi_{3p_z} = \Psi_{310} = \frac{2^{1/2}}{81(3\pi)^{1/2}}\left(\frac{Z}{a_o}\right)^{5/2}\left(6 - \frac{Zr}{a_o}\right) r\, e^{-Zr/3a_o} \cos\theta$$

$$\Psi_{3p_x} = \frac{2^{1/2}}{81(3\pi)^{1/2}}\left(\frac{Z}{a_o}\right)^{5/2}\left(6 - \frac{Zr}{a_o}\right) r\, e^{-Zr/3a_o} \sin\theta\cos\phi$$

$$\Psi_{3p_y} = \frac{2^{1/2}}{81(3\pi)^{1/2}}\left(\frac{Z}{a_o}\right)^{5/2}\left(6 - \frac{Zr}{a_o}\right) r\, e^{-Zr/3a_o} \sin\theta\sin\phi$$

$$\Psi_{3d_{xy}} = \frac{1}{81(2\pi)^{1/2}}\left(\frac{Z}{a_o}\right)^{7/2} r^2 e^{-Zr/3a_o} \sin^2\theta\sin 2\phi$$

$$\Psi_{3d_{xz}} = \frac{2^{1/2}}{81(2\pi)^{1/2}}\left(\frac{Z}{a_o}\right)^{7/2} r^2 e^{-Zr/3a_o} \sin\theta\cos\theta\cos\phi$$

$$\Psi_{3d_{yz}} = \frac{2^{1/2}}{81(2\pi)^{1/2}}\left(\frac{Z}{a_o}\right)^{7/2} r^2 e^{-Zr/3a_o} \sin\theta\cos\theta\sin\phi$$

$$\Psi_{3d_{x^2-y^2}} = \frac{1}{81(2\pi)^{1/2}}\left(\frac{Z}{a_o}\right)^{7/2} r^2 e^{-Zr/3a_o} \sin\theta\cos 2\phi$$

$$\Psi_{3d_{z^2}} = \frac{1}{81(6\pi)^{1/2}}\left(\frac{Z}{a_o}\right)^{7/2} r^2 e^{-Zr/3a_o} (3\cos^2\theta - 1)$$

TABLE A.7 The Radial Function, R(r), for Hydrogen-like Atoms

n	ℓ	R(r)
1	0	$R_{1s} = 2\left(\frac{Z}{a_o}\right)^{3/2} e^{-Zr/a_o}$
2	0	$R_{2s} = \left(\frac{Z}{2a_o}\right)^{3/2}\left(2 - \frac{Zr}{a_o}\right) e^{-Zr/2a_o}$
2	1	$R_{2p} = \frac{1}{\sqrt{3}}\left(\frac{Z}{2a_o}\right)^{3/2}\left(\frac{Zr}{a_o}\right) e^{-Zr/2a_o}$
3	0	$R_{3s} = \frac{2}{27}\left(\frac{Z}{3a_o}\right)^{3/2}\left[27 - \frac{18\,Zr}{a_o} + 2\left(\frac{Zr}{a_o}\right)^2\right] e^{-Zr/3a_o}$
3	1	$R_{3p} = \frac{1}{81\sqrt{3}}\left(\frac{2Z}{a_o}\right)^{3/2}\left(6 - \frac{Zr}{a_o}\right)\left(\frac{Zr}{a_o}\right) e^{-Zr/3a_o}$
3	2	$R_{3d} = \frac{1}{81\sqrt{15}}\left(\frac{2Z}{a_o}\right)^{3/2}\left(\frac{Zr}{a_o}\right)^2 e^{-Zr/3a_o}$